Foundations of Food Science

CW00621707

John Hawthorn
Professor of Food Science and
Head of Department of Food Science and Nutrition
University of Strathclyde

Foundations of Food Science

W. H. FREEMAN AND COMPANY
Oxford and San Francisco

W. H. Freeman and Company Limited

20 Beaumont Street, Oxford, OX1 2NQ
660 Market Street, San Francisco, California 94104

Library of Congress Cataloging in Publication Data
Hawthorn, John.
Foundations of food science.
Bibliography: p.
Includes index.
I. Food--Composition. I. Title.
TX551.H38 664 80-25623
ISBN 0-7167-1295-4 (U.S.)
ISBN 0-7167-1296-2 (U.S.:pbk.)

Copyright © 1981 W. H. Freeman & Co. Ltd.

No part of this book may be reproduced by any mechanical, photographic, or electronic process, or in the form of a phonographic recording, nor may it be stored in a retrieval system, transmitted, or otherwise copied for public or private use without the written permission of the publisher.

Typeset by The Universities Press, Belfast
Printed in the United States of America

Preface

Only once in his life does a food scientist have a chance to see his subject as a whole. Student days must therefore meet the double requirement of providing both a conspectus and an academic treatment of the subject which will lay the foundation for the intelligent choice of a future area of detailed expertise.

These objectives have to be achieved within the teaching time available in the undergraduate course. In a wide-ranging course this limitation precludes the possibility of approaching the frontiers of knowledge in all but limited areas. The problem is therefore one of balance and of selection. Opinions will differ on the balance to be achieved and on the material to be selected. If within these restrictions it is agreed that only the essential is relevant, there will still be controversy as to what is essential. Inevitably this book is one man's view.

To the Student

The book assumes that you, the student, have prior knowledge of microbiology and biochemistry amounting to about one hundred and fifty hours of lectures and laboratories in each subject, and attempts to lead you from there through the characteristics of the principal groups of human foodstuffs. Yet as you read and study, please remember the written word is no substitute for the face-to-face relationship of teacher and student. From this comes interest and motivation or boredom and failure. The tone of voice, the gesture and posture, the emphasis of conviction or the hesitation of inner doubt, add to the written word. A book may offer information but the teacher adds experience.

In using the book also remember that leisure is valuable. Effective study involves learning as much as possible in the minimum time. Work hard when you are working. Concentration is the secret of quick learning. Use the following rules.

1. *As you work, keep testing yourself for comprehension and recall*
   In addition to your lecture notes and text-book, keep a pocket-sized private notebook. Miss out easy connecting passages. Use this book to enter only the things you find difficult to understand or difficult to remember. Go over this once a week and check your recollection by covering the page and speaking *aloud* the passages to be remembered. Never work seriously against radio and television. For routine copying or other work requiring less concentration, a little light music is not

harmful as a background but never television. Whenever possible be in a room by yourself.

2. *Use lost time as study time, and it will not be wasted*
   The amount of time we spend every day waiting for something or other is surprising. Everybody misses buses or is caught in traffic jams. Carry your notebook. Instead of fuming with impatience use the time to check and recapitulate work. Stick formulae hard to remember by your shaving mirror. Use queuing time or cooking time or hair-washing and tooth-brushing time as remembering time. Time used this way is saved for dances or rugby or whatever is your pleasure. Practice doing this. You will be astonished how much time you can save.
3. *Start preparing for exams three weeks ahead*
   If you keep up with your work as described, three weeks is normally enough. In an exam you not only have to know the prescribed material, but you must also have it mentally organised so that you can use it. The best way to test this is by working over old exam papers, which can usually be obtained through your department or University or College library. This kind of practice done against the clock, teaches you what the examiners are looking for and how to organise examination time to make the best use of it. At the same time keep working hard on your notebook, checking and recapitulating the difficult bits.
4. *Don't work when you are mentally tired*
   Working till three or four o'clock in the morning may give a sense of virtue to the fool, but gives little else. It is a waste of time, unless you first sleep from eight till midnight. Believe it or not, but the human brain is at its best and most receptive just after sleep. If you are hard pressed, go to bed early, get yourself a cup of tea at 5 a.m. and put in two and a half hours at your books before breakfast. You will cover more work in this time than the five hours between 8 p.m. and 1 a.m.
5. *In lectures do not try to write down all the lecturer tells you*
   Pick up the key points, listen and concentrate on writing the bits you won't remember. If you do not understand, ask.

To the Teacher

Use this book as a background. Tell the class in advance the material you intend to cover in your assigned number of lectures. To the text give your own emphasis and illustrations, using film, transparencies, demonstrations and above all your own personal experience and anecdotes. The good teacher learns from the class and welcomes the question to which he does not know the answer. Officially you are teaching food science or food technology. If your first objective is to train your men and women to think as scientists and your second to direct their scientific attention to the study of foodstuffs you will not go far wrong.

Acknowledgements

To Professor John McLean who fired my eighteen-year-old imagination with the elegant delights of chemistry, to the late Mr. Hugh Clouston Moir who directed that interest to the chemistry of foodstuffs, to Dr. E. C. Bate Smith who widened my horizons to the multidisciplinary concept of food science and to the late Professor James Prior Todd who first gave me the chance to practise it, I owe my greatest debt. This book is their book. Only its faults are mine.

In the making of it, my thanks are due to Dr. K. Mary Clegg and Dr. R. C. C. Tao for corrections and criticisms of the chapter on nutrition, to Professor R. A. Lawrie for the improvements he suggested on the chapter on meat, to Dr. A. D. Ewart for advice on glutenin structures and to the publishers' referees for their helpful comments and criticisms.

The tables in the book were compiled from a wide range of sources. My aim has been to produce representative figures rather than to burden the student with the range of natural variation encountered in a given foodstuff. However, it is almost impossible to compile such tables without reference to McCance and Widdowson's data on food composition. The book owes a particular debt to successive editions of these tables, which now appear in updated form and beautifully revised by Dr. A. A. Paul and D. A. T. Southgate (H.M.S.O., London, 1978).

Finally I offer my sincere thanks to Mrs. E. Aubrey who typed much of the manuscript and to Mrs. M. Martin, my secretary since 1956, without whose loyal and unfailing help many of my endeavours, this one not the least, would have foundered.

March, 1980 JOHN HAWTHORN

Contents

Units and abbreviations

kilogram	kg	gram calorie	cal
gram	g	kilocalorie	kcal
milligram	mg	degree Celsius (centigrade)	°C
microgram	μg	degree Fahrenheit	°F
mole	mol	melting point	m.p.
metre	m	boiling point	b.p.
centimetre	cm	solution	soln
millimetre	mm	water activity	a_w
micro (10^{-6})	μ	hectare	ha
litre	l	adenosine monophosphate	AMP
millilitre	ml	adenosine diphosphate	ADP
microlitre	μl	adenosine triphosphate	ATP
hour	h	butylated hydroxyanisole	BHA
minute	min	butylated hydroxytoluene	BHT
second	s	coenzyme A	CoA
joule	J	inosine monophosphate	IMP
watt	W	inosine diphosphate	IDP
molar (moles per litre)	M	inosine triphosphate	ITP
millimolar	mM	free fatty acid	FFA
micromolar	μM	solids-not-fat	SNF

1

Human Nutrition

1.1. Introduction

Food science is simply the application of science to the study of foodstuffs. In its narrow sense it begins at a farm gate and ends on a dinner plate. It thus includes not only those characteristics of food which are subject to measurement and hence to scientific study and evaluation, but also the application of this knowledge to the problems of food processing. Urbanisation is impossible without the transport to and storage of food in the cities. Food must therefore be protected from spoilage and preserved by a wide range of techniques. Furthermore, in general, farm gate products are but raw materials. They must be converted by technology into the products we can use in home or in restaurant. If you don't believe me, try hand-milling your own wheat!

But this definition is too restrictive. The farm gate and the dinner plate concepts are barriers to the full implementation of the possibilities of the subject. In recent years progress in food technology has moved back through the farm gate to exercise influence on agriculture itself. To take a simple example, frozen peas provide freshness of flavour and texture twelve months of the year instead of the six to eight weeks of the normal pea harvest. But to obtain the best results, quick-freezing plants must use only selected varieties of peas grown on the right land properly prepared for the crop, harvested under peak conditions of ripeness (the crop will be at its best for only a few days), transported to the freezing plant in the shortest time after cutting, and frozen almost immediately under carefully specified conditions. To achieve all this, processors now specify in great detail to the farmer the preparation of the land, the type and quality of seed to be sown and when the crop has to be harvested. To obtain his co-operation they must also ensure that they offer a price for the crop which will reward him for the extra trouble and inconvenience involved in meeting their strict specification.

This example must suffice at this point but many others will become apparent as your studies proceed.

At the other end of the spectrum the food industry is now well aware of its duty to serve the nutritional needs of the final consumer. Not so long ago and well within the working experience of many food scientists now in their forties or over, most food companies considered that their prime job was to manufacture and sell foodstuffs which were attractive to look at, enjoyable to eat and satisfying to the appetite. But appetite in itself does not prevent people from eating ill-balanced meals. Furthermore, the knowledge of the relationship between food and health is no longer a field reserved to the nutritionist or the doctor. Helped by radio, television, and often excellent popular articles in the newspapers and magazines, a rapidly growing army of housewives (and husbands too) is demanding more information about the nutritional qualities of what they eat and food processors neglecting this interest and concern do so at their financial peril.

We now know that many of the ills to which mankind has always been exposed can be completely avoided by proper nutrition, that health and vitality demand the right foods, that choosing a well-balanced diet is not difficult nor need it be expensive, and that good eating habits are the best defence open to the ordinary man and woman against disease and the best guarantee of a long, active and virile life (let the ladies excuse the word virile because in this sense it applies to them too) and for a healthy and tranquil old age.

Thus while food scientists do not claim to be practising nutritionists they must have a working knowledge of the basic principles of nutrition and of nutritional processes. They cannot otherwise discharge their day-to-day responsibilities. If they extend their principles to their personal eating habits they set a practical example to their fellows. As I write I think of three food scientists or nutritionists who are personal friends, age range 76 to 82 years; all are still professionally active, all have led busy and full lives and all three are fitter than many men twenty-five years their juniors. Shows what a good constitution coupled with good nutrition can do for you!

Before beginning the study of foodstuffs, we must start this book by outlining the basic principles of nutrition. They will underpin all our subsequent considerations.

1.2. Water and minerals

The human body is contrived of materials obtained from the food it consumes in conjunction with the air it breathes. The first step in human nutrition is to consider the gross composition of our bodies. A typical human male weighs about 70 kg, and his body composition approximates to

Water	43 kg
Protein	12 kg
Fat	11 kg
Minerals	4 kg

At a glance it can be seen that we are more water (62 per cent) than anything else. A loss of 2 kg of this water causes discomfort, 4 kg is disabling and death follows rapidly at 8 kg. Our water requirements per day depend on the water lost in perspiration, evaporation from the internal surfaces of the lung as water vapour exhaled in breath, and in urine. Under ideal conditions of rest and cool surroundings the minimum water requirement is 1 kg per day and this is mainly required to compensate for urinary losses. In practice we almost always require much more than this, and an athlete in training or a man doing hard physical work in the tropics can lose 4 kg per day in sweat alone, and this must be replaced.

As well as intake by drinking, we obtain useful quantities of water from food. Most raw vegetables and fruits commonly consumed in European countries contain 75 per cent water or more (although this does not apply to dried fruits such as dates, raisins and prunes) and many contain over 90 per cent. Most meat and fish dishes as eaten will contain at least 50 per cent water, and even bread, which we regard as a dry food, contains between 35 and 40 per cent. Fresh whole cow's milk, which is often and correctly thought of as a rich and valuable food, contains about 87 per cent water and lettuce tops the list with chicory at around 95–96 per cent.

Our typical 70-kg man, if reasonably active, would eat about 1.5 kg of food each day of which about 1 kg would be water contained in the food. Of course, these are only rough figures but they do remind us that an appreciable fraction of our water intake comes from the food we eat. They also remind us that ($1.5 \times 365 = 548$ kg) most people eat something over half a tonne of food, or seven to eight times their own body weight each year.

The 40 kg of water provides the medium in which the near-magic of human biochemistry takes place. The 12 kg of protein provides the plumbing and muscular power units and much of the control structure housed in the skull and called the brain. This protein is in a continuous state of breakdown due to what can roughly be described as normal wear and tear, and in the event of injury or illness the breakdown rate is increased. This loss must be replaced by food protein, about which more due course.

The fat is mainly located as insulation under the skin and in the abdominal cavity as a reserve store of energy, but some is present in the form of more complex lipid materials, which have functional and structural properties essential to the proper functioning of certain bodily tissues and processes. Of the 11 kg of fatty material present, probably not more than 1 kg is essential to the proper working of the biochemical machinery, the remainder being present in a convenient form for mobilisation during the period of enforced fasting which so often must have been the lot of human-kind before the development of agriculture. A man carrying 5 kg of excess fat is carrying about 12 days reserve food (energy) supply. This is clearly a useful thing to do if you are likely to be faced with a food shortage. In well-fed societies, excess weight, which tends to increase with age and is never pruned by famine, is a major health hazard.

Of the 4 kg of minerals, about 3 kg is made up of calcium phosphate which gives rigidity to bone structures. Of this, about 1.2 kg is calcium, which is thus the dominant metal of body construction and one which also plays a key role in intimate processes such as muscle contraction. The associated phosphate is the equivalent of around 600–900 g of phosphorus. Since phosphate is a constituent of all plant and animal cells, it is almost ubiquitous in foodstuffs, except in highly refined products such as sugar. Dietary deficiency of phosphate is not known in man.

Following calcium, the next most important is potassium, which is present in the form of about 285 g of potassium chloride. Sodium follows closely behind at the equivalent of 250 g of sodium chloride. Thereafter the amounts of metals essential to life processes taper off. Magnesium is required at a level of 25 g. Despite its key role in blood and muscle function, the whole body iron content is only around 4 g or enough to make a two-inch nail. Zinc is present in at least eight enzyme systems and adds up to a total body burden of 1.5 g. Copper is even less at around 0.15 g, manganese is also present in small amounts while cobalt, although critically essential to health, is only present at levels of less than 1 mg. Molybdenum is also required in small amounts for the functioning of the xanthine oxidase enzyme system. Amongst the non-metals, iodine is found in tiny quantities, the total body burden being about 40 mg, and fluorine is regarded as being essential (in trace amounts) to dental health in children. Apart from the nitrogen and sulphur present in the structure of proteins this completes the short list of elements essential to human life. Of course other substances have been detected in human tissues but, so far as present knowledge goes, they are intruders accidentally introduced into our bodies by their presence in food and serving no known function.

Selenium, chromium, cadmium, lead, mercury, arsenic, lithium, strontium, boron, tin, vanadium, nickel, silicon and aluminium are all to be found in trace quantities in human tissues, but there is no evidence that deficiency of any of them is responsible for any disorder of man. However, some appear to be essential nutrients for farm animals and may therefore play as yet unidentified parts in human metabolism. It should be noted that in large amounts selenium, cadmium, lead, mercury and arsenic are toxic metals. The presence of extraneous substances is not surprising. Plants absorb minerals from the soil both actively, in the sense that they are required by the plant for its metabolism or structure, and passively simply because they are soluble in the water which is taken up by the plant roots. Since plants are eaten by man and form the food of his domestic animals, mineral elements pass through this route into the human gut, and some of them thereafter find their way into various tissues even if they have no specific metabolic role to play.

Having looked at the constituents of our bodies it is clear that, for practical purposes, they must all be derived from the food we eat, and that is what the science of nutrition is all about.

1.3 Energy metabolism

The energy required to maintain our bodies at their working temperature of 37°C must be obtained by oxidation of food constituents. Surplus protein may be disposed of in this way and yields about 4 kcal/g (16.7 kJ/g) on combustion through bodily metabolic processes. For practical purposes carbohydrate gives about the same figure (3.8 kcal/g; 15.9 kJ/g) while fat gives 9 kcal/g (37.6 kJ/g), and alcohol 7 kcal/g (29.3 kJ/g). These factors allow for availability and losses of combustible material in faeces and urine and, although subject to criticism in some respects, they are accurate enough for most practical nutritional calculations and should be memorised.

Since the different forms of energy are interconvertible, energy used for basal metabolism (that is to say for pumping blood, heart and lungs and for keeping the body warm) can be as conveniently considered in kcal as the mechanical energy required for physical movement. On the recommendations of the International Union of Nutritional Sciences, there has been an attempt with other official backing to replace the kcal (which is a unit of heat) with the joule (J) which is a unit of energy, the joule being defined as the energy expended when 1 kg is moved 1 meter (m) by a force of 1 newton (N). The purpose behind this change is to make nutritional units consistent with S.I. (*Système International*) units of measurement now widely accepted throughout the world by the physical sciences. The joule is too small a unit for convenient use in nutrition and food science and in practice the kilojoule (kJ; equal to 10^3 J), or the megajoule (MJ; equal to 10^6 J) is preferred. Kilocalories can be converted to kilojoules by multiplying by 4.184. For most practical purposes a factor of 4.2 is sufficiently accurate. Although several years have passed since these changes were proposed, they have gained only limited acceptance in the more learned journals. In popular use the kilocalorie still dominates and therefore both units will be given in this book. In metabolic processes, the greater part of the energy metabolised appears ultimately as body heat, and familiarity apart, this suggests an additional reason for retaining the older unit.

For the past fifty years popular writers on nutrition all round the world have attempted to familiarise their readers with the concepts of food energy embraced in the word 'Calorie', representing the former kcal of physics. They have largely succeeded. Think of the position of a local nurse in a bush village in Ghana trying to explain to the mother of a malnourished child in Ga, one of the local languages, the difference between Calories and Joules as laid down in the *Système International* and as endorsed by the Royal Society in London. The term 'Calorie' is firmly entrenched in popular usage and may well remain so entrenched for decades to come. Modern food scientists will no doubt wish to use kilojoules, but all students of food science should be aware of the grass roots dilemma which this change has imposed on dietitians, nurses, clinicians, popular journalists and ordinary housewives. Nutrition is not merely a matter for nutritional scientists.

Measuring human energy expenditure is simple in theory but rather difficult and costly in practice. The classical experiments were carried out by Atwater and his colleagues. They constructed a small heavily insulated room sufficiently large for a man to be seated in comfort and to carry out simple tasks. The heat produced in the chamber was removed by water coils and the amount produced was calculated from the rate of water flow through the coils and the difference between the inlet and outlet water temperatures. Arrangements were made for supplying the subject with circulating air passed through sulphuric acid (to absorb the water produced by transpiration) and then through soda lime (to absorb carbon dioxide). The oxygen removed was replaced by pure oxygen from a cylinder. A window was provided through which the subject could be observed, and a port-hole was used to remove excreta and to provide food.

Using this type of apparatus it has been demonstrated (as would, of course, be expected) that the fundamental physical law of conservation of energy applies equally to men and machines. In the human case the total energy expenditure, which of course is the sum of the mechanical work done and heat produced, was found to be the same as the chemical energy in the food consumed less the energy lost in the faeces and urine. It was also found that total energy expenditure is quantitatively related to the oxygen consumed. This second finding greatly simplified subsequent work, because it made possible the use of indirect calorimetry (which is merely a measurement of human oxygen consumption) for measuring human energy outputs. Using this one can measure a much wider range of energy expenditure in human activities, from the worker at the coal face and the lumberjack wielding his axe to the energy expenditure during sleep—and all this without shutting the experimental subjects in an air-tight insulated box.

From this kind of experiment, work can be categorised in terms of the human energy required to perform it. Thus light assembly work in industry, many kinds of housework, military drill, painting, playing golf and driving a truck involve energy expenditures of from 2.5 to about 5.0 kcal/min (10–20 kJ/min); a labourer using a pick and shovel, a soldier on a route march with rifle and pack, or a subject playing tennis or cycling at speeds of up to 10 m.p.h. requires an output of from 5 to 7.4 kcal/min (20–31 kJ/min); heavy work, such as coal mining or playing football takes the output up to 10 kcal/min (40 kJ/min); and very heavy work, such as swimming crawl, working as a furnace man in the steel industry, working as a lumberjack or cross-country running may result in outputs of over 10 kcal/min.

As mentioned earlier, basal metabolic rate refers to the energy requirement of the body at rest, and it is the energy needed for the activity of the internal organs and to maintain the body temperature at 37°C. This figure is high in young, rapidly growing children but stabilises after about the age of twelve years and thereafter very gradually declines into old age. Basic metabolism is similar in different species of animals when calculated on the basis of body surface area. For humans and on the same basis, men have a higher basal metabolic rate than women, but this reflects the fact that women are fatter than men and have a larger surface area per unit of lean body mass. Calculated on a basis of lean body weight, the rates for both sexes are much the same. Quoting average figures, and age for age, the fat content of the human body (expressed as a percentage of body weight) is about twice as great in women as in men although there is a tendency for the relative difference to diminish with age.

If energy intake, whether expressed as calories or joules, exceeds energy expenditure (physical work plus heat generated) the surplus is deposited in the form of body fat. This rule applies whatever the form of the surplus, be it protein, fat or carbohydrate. In this sense no foods are specifically fattening. The idea that some foods are 'slimming' foods is equally fallacious. If energy expenditure exceeds energy intake, loss of weight occurs in mammals. This is as inexorable as the law of conservation of energy in physics with some very rare and only apparent exceptions where excessive retention of body water in some diseased states may produce apparent anomalies. In the preponderance of cases, overweight is the result of eating in excess of bodily needs.

Unfortunately there is a tendency to associate obesity with greed and lack of self-control. This is not necessarily true. The amount an individual eats is controlled by appetite, the mechanism of which is as yet not thoroughly understood although there seems little doubt that it is partly regulated by the hypothalamus, an organ situated at the base of the brain. It seems that both physiological and psychological factors are involved and a first step in treating obesity is the realisation that it is a condition deserving sympathy rather than blame. It is outwith the scope of this book to discuss its treatment, but it is necessary to state firmly that obesity is an important health hazard which reduces life expectancy, increases susceptibility to circulatory complaints ranging from varicose veins to diabetes and coronary artery disease. In general terms it is one of the most serious health hazards of modern urban communities and in its various manifestations it should be regarded as a disease and a killer disease at that.

Having said this, it is worth recalling that the ability to metabolise excess energy and lay it down as fat was an advantage to our hunter-gatherer ancestors. For nearly two million years our forebears led lives of uncertainty so far as the next meal was concerned. The fat individual carried his personal food reserves and could survive a famine better than his lean neighbour. Perhaps this is why to this day women survive exposure and food deprivation better than men (remember their higher fat content per unit of body weight) and this in turn may hinge on the greater need for femine survival for the continuance of the species. Only when the advent of agriculture made food supplies more secure, did obesity become a hazard rather than an advantage. However, hazard it certainly is for the three-meals-a-day society, especially when coupled with lack of physical exercise and the stresses of modern living.

In the previous section some general figures were quoted for energy outputs per hour in different kinds of activities. A large man will have greater basal metabolic requirements than a small man and must therefore eat more. Women, being smaller and better insulated than men, require less food energy on an age-for-age basis. Average energy requirements in the male increase from birth up to the age of about twenty-two and thereafter slowly decline until in their fifties their daily energy requirement will be 80 per cent of that of their early twenties. Body fat content tends to increase with age and this may reflect a relatively constant food energy intake against a declining food requirement.

The patterns of energy requirement for women are distinctly different. Up to the age of eleven years their needs are similar to those of their male counterparts, but reach a peak at the age of fourteen. Thereafter they gradually decline until in their fifties their energy needs have dropped to 75 per cent of the requirement at fourteen. However, during pregnancy energy needs rise by 300 kcal/day (1260 kJ/day) rising further to an additional 500 kcal/day (2100 kJ/day) during lactation.

In almost all human food patterns carbohydrates are the major source of energy, and for poor peoples (especially in tropical areas) up to 90 per cent of the diet may be carbohydrate based. Since a high carbohydrate diet is monotonous, carbohydrate levels tend to decline as income levels rise. Even in wealthy communities, staple foods tend to be high in carbohydrate and total intakes are likely to exceed 50 per cent of the total calorie requirement.

While the number of known sugars and other carbohydrates is large, those of quantitative significance in human diet are quite small. Glucose is of prime interest as the only hexose known to exist in the free state in the fasting human body, being always present in blood at a level of around 1 mg/ml. It is also stored in muscle and liver in the form of its readily mobilised water-soluble polymer, glycogen. Muscle glycogen levels are depleted during violent exercise but are readily replaced from the circulating pool of blood glucose. Sucrose in the form of cane or beet sugar is consumed in quantity in European and North American diets, the average intake in the United Kingdom being of the order of 56 kg per person per annum or, say, 570 kcal (2400 kJ) per person per day. In round figures this represents about 21 per cent of the total energy intake. Since high sucrose consumption has been implicated in a number of human ailments ranging from dental caries to high blood cholesterol levels, some concern has been expressed in nutritional circles at these figures. Perhaps the most serious criticism is that of 'empty calories'. Most high-energy foods, such as cereals, also contribute other nutrients such as vitamins and minerals. As sold, sucrose is a highly purified chemical which contributes nothing but energy. Taken as a substantial fraction of the total diet, it displaces other energy sources endowed with supplementary nutrients thus lowering the overall quality of the diet.

Lactose is the disaccharide unique to mammalian milk and derived from glucose and galactose. Human milk contains much more lactose and less protein than cow's milk and cow's milk fed to human infants is often modified by dilution with water and rebalanced with added sucrose. Infants and young children occasonally show an intolerance to lactose characterised by fermentation in the large intestine and the production of loose stools containing lactic acid. Lactose intolerance appears to be much more common in Africa and some parts of Asia than in Europe. It is due to defective production of the enzyme lactase in the small intestine. Maltose is a disaccharide derived from two linked molecules of glucose. It is formed from the breakdown of starch in the malting of cereals. It does not have special nutritional significance so far as humans are concerned but it is of great industrial importance to the fermentation industries.

Fructose, a monosaccharide has the same molecular formula ($C_6H_{12}O_6$) as glucose but with a different structural formula. It is found in free form in some fruits and in honey. Since it is laevorotatory it is sometimes known by its alternative name of laevulose.

Ribose and deoxyribose are pentose sugars present in small amounts in all living cells as constituents of the nucleic acids and of high-energy compounds such as adenosine triphosphate. Ribose can be synthesised by animals and man and is not an essential dietary constituent nor is it a significant source of dietary energy. However, two other pentose sugars, arabinose and xylose, are fairly widely distributed in fruits and roots and are normal dietary constituents.

Forgetting is easier than remembering. The student is advised to recheck his memory of sugar structures from the formulae below:

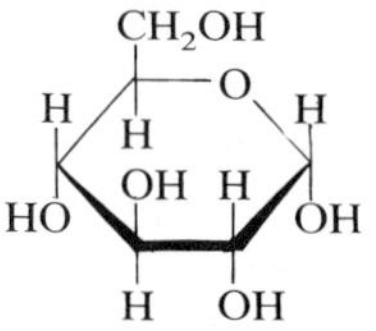

α-D-Glucopyranose 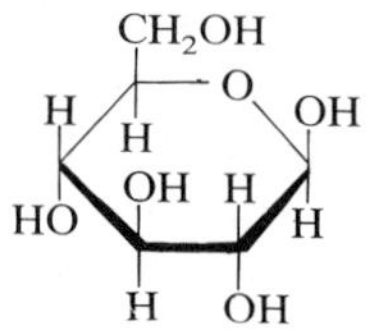 β-D-Glucopyranose 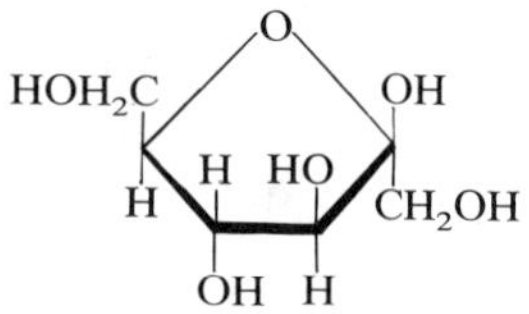 β-D-Fructofuranose

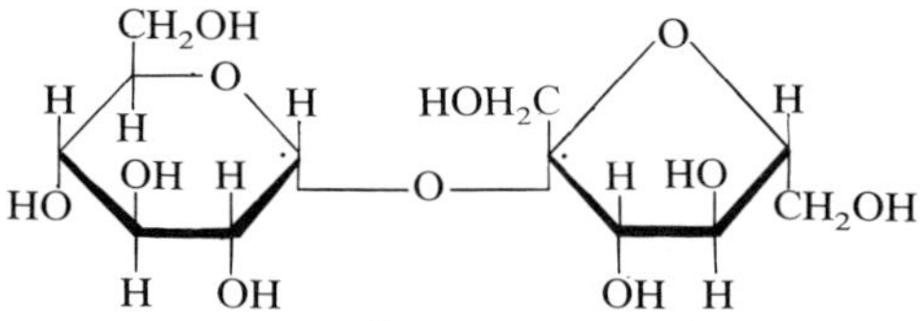

Sucrose

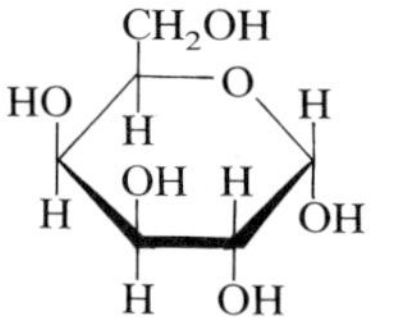

α-D-Galactopyranose 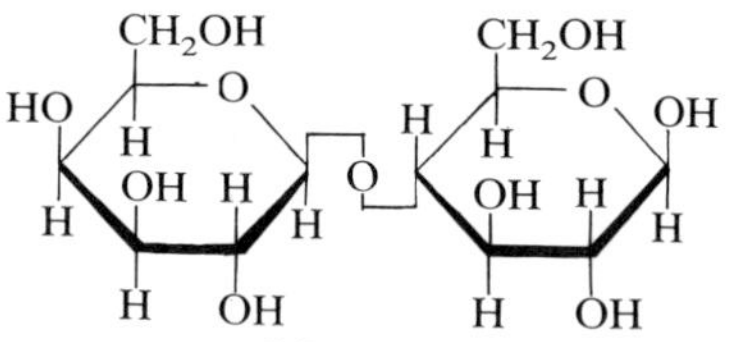 β-Lactose

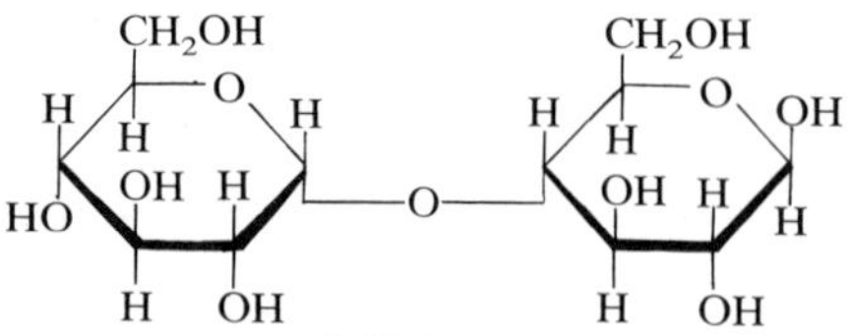

β-Maltose

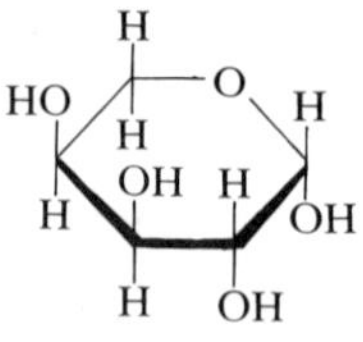

α-D-Arabinopyranose 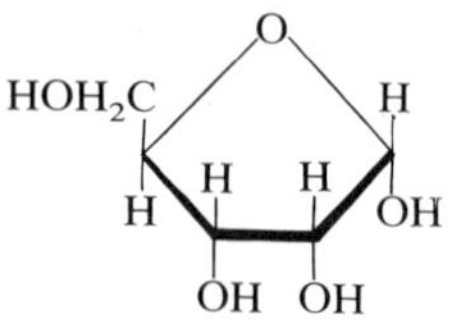 α-D-Ribofuranose 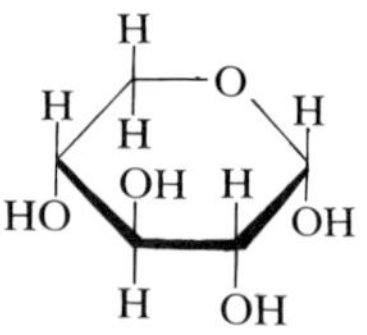 α-D-Xylopyranose

Starch is the major energy source in human diet, and occurs in granules whose size and shape are characteristic of the plant from which it is obtained. Intact starch granules are insoluble in cold water but the membrane is ruptured by heating, to provide a viscous solution or a gel depending on the concentration used. In general starch consists of two fractions known as amylose and amylopectin. On treatment with acid or enzymes it is progressively degraded to dextrin (a mixture of low molecular weight polysaccharides), maltose and finally to glucose. Commercial corn syrups normally contain all three. Amylose consists of at least two hundred glucose units linked together through α1–4 links as follows:

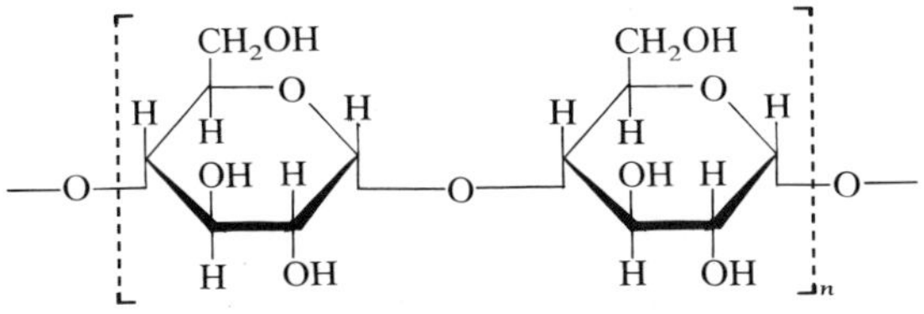

Repeating maltose unit of starch

The molecular weight is around 30,000–40,000. In contrast, amylopectin has a molecular weight of around five times these values, yet chain length determination by end-group analysis indicates a chain length averaging twenty units. The chains must therefore be branched. Branching takes place through α1–6 links. Starch is readily available for digestion in the human gut and is absorbed in the form of glucose.

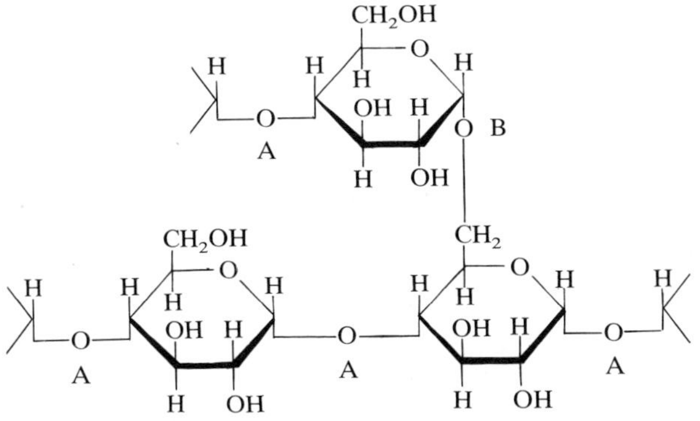

Branch point of amylopectin and glycogen
$A = (\alpha 1\text{–}4)$ glucosidic bond
$B = (\alpha 1\text{–}6)$ glucosidic bond

Cellulose is the chief component of plant fibres, is insoluble in water and of extremely high molecular weight. It is scarcely, if at all, digested by man, although it is a major energy source for ruminant animals whose digestive processes are assisted by the microbial flora of the rumen. For long regarded as unimportant in human diet, evidence is now accumulating to indicate that its insolubility may have an important physical function in the processes of the large intestine, where it resists over-dehydration (and therefore compaction) of the faeces and provides for more bulky stools facilitating the peristaltic muscular movements which eventually lead to elimination. Primitive diets tend to be much higher in cellulose than those of urban societies and some of the so-called diseases of civilisation affecting digestive processes may not be unrelated to this. The recurrent public controversy concerning the use of white breads and refined sugars in part hinges on the resulting low cellulose levels in the digestive system. Pectic substances and hemicelluloses are also important in this context.

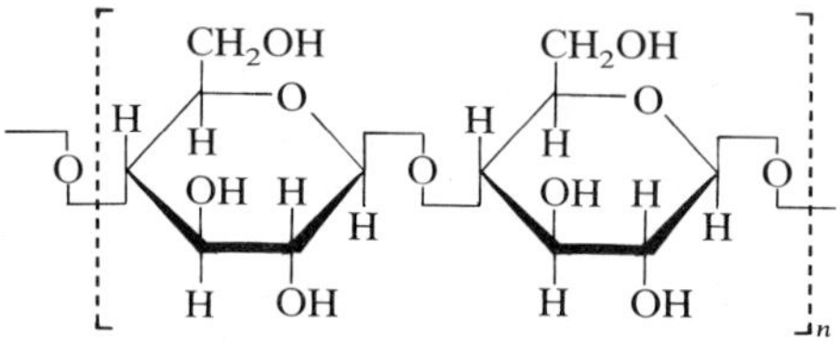

Repeating cellobiose unit of cellulose

In cellulose, the repeating glucose units are joined as shown above through β1–4 linkages characteristic of the disaccharide cellobiose.

1.4. Lipids

The terms 'oils' and 'fats' are clear in the mind of someone purchasing groceries, but inexact to the chemist, who generally prefers to restrict their use to describe neutral triglycerides. He prefers the general word 'lipid' to describe fat-related substances whether fatty acids, neutral or partial triglycerides, waxes, phospholipids or sterols.

To the nutritionist it is useful to distinguish between fats in the sense of butter, margarine and cooking fats and the lipid materials which are present in less obvious ways. For example, many samples of foods which are considered to be non-fatty contain appreciable quantities of lipid. Thus white flour may contain only up to 2 per cent lipid but this represents 5 per cent of the purchased calories. Lean beef, trimmed as clear of fat as possible (the popular concept of a protein food) will normally contain around 30 per cent fat and 60 per cent protein on a dry-weight basis.

Since the fat content of food as purchased is variable and since the processes of preparing and cooking will add further fat depending on the cook, and since the consumer, for example, may well put butter on his toast but leave his bacon fat trimmings on the edge of his plate, it can be difficult to ascertain the true fat content of a given diet by calculation from nutritional tables, and resort has to be made to direct chemical analysis.

When a person becomes obese the fat cells of adipose tissue enlarge as additional fat is deposited. Excessive fat deposits are not necessarily evenly distributed over the principal fat depots. These are (a) immediately under the skin and often particularly in the region of the abdomen and thighs, (b) in tissues surrounding the kidney and the heart, (c) between the muscles, referred to as intermuscular fat and (d) within the muscles themselves, the fat being deposited in the collagenous layers separating the bundles of muscle fibres. In starvation the depots surrounding the heart and kidney are normally the last to be depleted.

Since there are few communities in which fat does not provide at least 20 per cent of the dietary energy and in many the figure is well over 35 per cent, no accurate estimates of minimum fat requirements for humans exist. Volunteers have lived for periods of up to six months on diets virtually fat-free without obvious ill-effects and it seems likely that physiological requirements for fat are very much lower than are likely to be encountered in practical situations. However, it is known that for many animals and insects, growth and survival is dependent on the dietary presence of small amounts of what are known as essential fatty acids—linoleic, linolenic and arachidonic acids. Of these arachidonic is probably the most physiologically active form but since the others can readily be converted to it in the body, the distinction may be academic. The amounts required are not known with certainty since direct proof that dietary deficiency of essential fatty acids exists in man is lacking. In view of the indirect evidence of animal experiments it is prudent to assume that they are required. However, the amounts required are likely to be small, probably not more than 1 g/day. In middle and later life atheroma in all its ramifications and consequences may be exacerbated by a high intake of saturated fats, and there are indications that replacing all or part of these with highly unsaturated fats may be beneficial.

The following structures should be memorised as they are frequently referred to in the study of food science.

Saturated fatty acids:

$C_{11}H_{25}$—COOH *n*-dodecanoic (lauric) acid
$C_{13}H_{27}$—COOH *n*-tetradecanoic (myristic) acid
$C_{15}H_{31}$—COOH *n*-hexadecanoic (palmitic) acid
$C_{17}H_{35}$—COOH *n*-octadecanoic (stearic) acid
$C_{19}H_{39}$—COOH *n*-eicosanoic (arachidic) acid.

Unsaturated fatty acids:

$C_{15}H_{29}$—COOH	*cis*-9-hexadecanoic (palmitoleic) acid
$C_{17}H_{33}$—COOH	*cis*-9-octadecenoic (oleic) acid
$C_{17}H_{31}$—COOH	*cis*-*cis*-9,12-octadecadienoic (linoleic) acid
$C_{17}H_{29}$—COOH	all-*cis*-9,12,15-octadecatrienoic (linolenic) acid
$C_{19}H_{31}$—COOH	all-*cis*-5,8,11,14-eicosatetraenoic (arachidonic) acid.

If the general formula of fatty acids be considered as R—COOH, then neutral triglycerides, which form the major component of naturally occurring fats, will have the ester formula, where R_1, R_2 and R_3 may be the

$$\begin{array}{l} CH_2—O—\overset{O}{\overset{\|}{C}}—R_1 \\ | \\ CH—O—\overset{O}{\overset{\|}{C}}—R_2 \\ | \\ CH_2—O—\overset{O}{\overset{\|}{C}}—R_3 \end{array}$$

same or different fatty acid residues. If one fatty acid is replaced by a hydrogen atom the resulting lipid is referred to as a diglyceride and if two of the three are hydrogens, the substance is a monoglyceride. Since the esterified glycerol molecule may be made asymmetric by esterifying the 1 and 3 carbons with different fatty acids, the 1 and 3 positions on the molecule are not exactly equivalent as the Fisher projection would suggest. Thus many natural glycerides show a stereospecific distribution of fatty acids rather than a random or 1,3 random one. For example, the characteristic short-chain fatty acids of milk fats tend to accumulate in position 3, while animal depot fats tend to have saturated fatty acids in position. 1. This non-random effect is of more interest to the structural biochemist than to the practical study of nutrition, but it is also of interest to the food scientist concerned with fat technology.

The phospholipids include a large number of substances characterised by having a fatty acid chain coupled to a phosphate group either through glycerol or sphingosine.

$$\begin{array}{l} \qquad\qquad\quad CH_2—O—\overset{O}{\overset{\|}{C}}—R \\ \qquad\qquad\quad | \\ R_2—\overset{O}{\overset{\|}{C}}—O—CH \\ \qquad\qquad\quad | \\ \qquad\qquad\quad CH_2—O—\overset{O}{\overset{\|}{P}}—O—X \\ \qquad\qquad\qquad\qquad\;\; | \\ \qquad\qquad\qquad\qquad\; OH \end{array}$$

Glycerophospholipids

$$\begin{array}{l} CH_3—(CH_2)_{12}—CH{=}CH \\ \qquad\qquad\qquad\qquad\quad | \\ \qquad\qquad\qquad HO—CH—CH—CH_2—O—\overset{O}{\overset{\|}{P}}—O—X \\ \qquad\qquad\qquad\qquad\qquad\quad | \qquad\qquad\qquad\;\; | \\ \qquad\qquad\qquad\qquad\qquad NH \qquad\qquad\quad OH \\ \qquad\qquad\qquad\qquad\qquad\quad | \\ \qquad\qquad\qquad\qquad\qquad\; \underset{\underset{O}{\|}}{C}—R \end{array}$$

Sphingophospholipids

In these phosphatides the phosphoric acid is also bound through an ester linkage to one of the following nitrogeneous compounds in the position shown above as X.

$HO{-}CH_2{-}CH_2\overset{+}{N}{\equiv}(CH_3)_3$
Choline

$HO{-}CH_2{-}CH_2{-}NH_2$
Ethanolamine

$HO{-}CH_2{-}CH(NH_2){-}COOH$
L-Serine

The possible permutations of this pattern include a very large group of naturally occurring substances. In addition to those illustrated, links based on inositol, glucose and galactose are also found. By far the most abundant of the phospholipids are those based on glycerol and examples are found in every living cell.

The sterols and waxes (fatty acid esters of higher alcohols) are also grouped with the lipids, and the sterols in particular will be considered elewhere.

The absorption and metabolism of lipids (as of carbohydrates) is outwith the direct scope of this work and for more detailed information the reader must look to texts on physiology and biochemistry. However, four principal points should be remembered. (1) The end-products of the absorption and metabolism of carbohydrates and lipids are energy plus carbon dioxide and water. (2) An intake of either of these in excess of the body's energy requirement will be stored, ultimately in the form of adipose tissue. (3) On a weight-for-weight basis, lipids deliver more than twice as much energy as carbohydrates. (4) This apparently simple equivalence does not necessarily mean that, calorie for calorie, carbohydrates and lipids are equally satisfying to the person on a slimming diet.

1.5. Amino acids and proteins

It is a matter for philosophical reflection that the same twenty amino acids occur in all terrestrial life forms, whether animals, plants or micro-organisms. A few other amino acids of limited distribution are also found but occur in only a very few kinds of protein. There is a unity which underlies the surface diversity of living forms. The food scientist is duly grateful for anything which permits him simplifying assumptions.

The formation of peptides and proteins from amino acids is dependent on the peptide bond which permits the stringing together of two, three or many amino acids of the same or of different structures to form proteins. For example, bovine insulin is a comparatively simple protein of fifty-one amino acid residues and a molecular weight of around 5,700 but most proteins are of much higher molecular weight than this and figures ranging up to nearly 3,000,000 have been reported. Thus five thousand

and more amino acid residues are commonly encountered in a single protein molecule.

The English alphabet has twenty-six letters and a well-educated Englishman will probably use up to ten thousand words constructed from these letters. Since only a small proportion of these words contain more than twelve letters and since there are certain restrictions as to the letter sequences that are used in practice, it is clear that almost the whole range of the human response to the environment can be handled through these twenty-six structural elements. The twenty amino acids offer similar scope for variety but when it is remembered that each 'word' in the protein scheme consists of at least forty letters and will more often consist of many times this figure, the chemical scope for producing diverse products becomes clearer. For practical purposes the range of protein structures is unlimited, and nature puts this diversity to good use in the protein products she deploys in the plants and animals we use for food.

Of the twenty amino acids, eight must be obtained directly from food since they cannot be synthesised *in vivo* at a rate to meet metabolic requirements. These are

tryptophan	to
methionine	make
lysine	love
threonine	to
leucine	Lucy
isoleucine	is
phenylalanine	positively
valine	voluptuous

Histidine is also required during the rapid growth period of human infancy and arginine is sometimes also classified as an essential amino acid because while it can be synthesised (for example by the rat) the rate of synthesis is too low to permit optimum growth. The student should memorise this list (hence the mnemonic) and the corresponding structures below

$C—CH_2—CH(NH_2)—COOH$ ‖ CH N (benzene ring fused)	tryptophan
$CH_3—S—CH_2—CH_2—CH(NH_2)—COOH$	methionine
$CH_2(NH_2)—CH_2—CH_2—CH_2—CH(NH_2)—COOH$	lysine
$CH_3—CH(OH)—CH(NH_2)—COOH$	threonine
$CH_3—CH(CH_3)—CH_2—CH(NH_2)—COOH$	leucine
$CH_3—CH_2—CH(CH_3)—CH(NH_2)—COOH$	isoleucine

C_6H_5—CH_2—$CH(NH_2)$—COOH phenylalanine

CH_3—$CH(CH_3)$—$CH(NH_2)$—COOH valine

A casual examination of these structures points to the generalised formula R—$CH(NH_2)$COOH to describe the amino acids. They can be linked together through the basic amino group of one amino acid to the acidic carboxyl group of another with the elimination of one molecule of water

$$R\text{—}CH(NH_2)\text{—}CO[OH+H]NH\text{—}\underset{COOH}{\overset{R}{C}}\text{—}H \rightarrow R\text{—}CH(NH_2)\text{—}CO\text{—}NH\text{—}CH(R)\text{—}COOH+H_2O$$

When the reaction is complete the new double amino acid (a dipeptide) still has an amino group and a carboxyl group free to react with further amino acids of the same or of different structures, and this process of condensation produces first polypeptides and finally proteins.

The digestion of food proteins is roughly this process in reverse. In the stomach is secreted the enzyme pepsin which works best at around pH 1–2. By a quite remarkable feat of localised chemistry this is provided in the stomach by the release of hydrogen and chloride ions which effectively produce hydrochloric acid. Pepsin breaks down proteins to smaller polypeptide units, which in due course pass on to the upper part of the small intestine where they are mixed with trypsin, the chief proteolytic enzyme of pancreatic juice. In contrast to pepsin, trypsin prefers an alkaline medium of pH 8 and the digestion of proteins is continued until they are broken down to amino acids and small peptides, which are then absorbed through the intestinal wall and into the blood. There is good evidence that peptides containing up to six amino acids may be absorbed as rapidly as the free amino acids.

The absorbed amino acids may be used in three ways: for tissue growth, for replacement of lost proteins (tissue maintenance) or for energy. The rapid growth periods of infancy and childhood require large protein intakes per unit of body weight. For example, according to the recommendations of the United Kingdom Department of Health and Social Security, during the first year of life an infant should receive 20 g of protein per day. An active middle-aged adult should receive only about three times that amount despite the fact that his or her body weight will be about ten times that of the infant.

At all times of life, protein is lost daily in sweat, from the mucus membranes and from faeces. There may be further losses by bleeding from minor or major injury or from menstruation. In addition, proteins are deaminated and oxidised to provide energy in much the same way as carbohydrates and fats, the nitrogen released being excreted in urine as urea. These are short-term mechanisms and will vary in extent with external factors such as the amount of food consumed, the amino acid spectrum of the food protein and the amount of exercise taken. Most of our information comes from balance studies in which nitrogen intake is carefully measured and compared with nitrogen excreted from all sources. If the intake exceeds the output, the subject is retaining a proportion of the protein consumed and is considered to be in positive balance. The reverse situation in which the subject is excreting more than is consumed is described as negative nitrogen balance. Much of our knowledge of human protein requirements comes from such experiments. However, they are tedious to carry out (for the experimenter and the subject), disrupt the subject's normal life pattern and thus are normally conducted over short periods of time. We do not really know very much about whether nutritional need for proteins varies from time to time in the healthy adult, although the need does increase during pregnancy and lactation.

As a starting point it seems sensible to assume that protein requirement is related to body weight in adults, and much observation and experiment has tended to confirm that this is a reasonable approximation which works well enough for all practical purposes. It is generally also assumed that those engaged in heavy work need rather more protein than more sedentary men and women, but it has proved difficult in practice to determine protein requirements with any precision. There are two main reasons for this. The first is that food proteins differ in their efficiency of utilisation by the body when fed as single protein sources. Thus, children absorb and utilise milk and egg protein more effectively than maize and rice protein when these are taken singly. In general, animal and fish proteins are 'better' in this sense than vegetable proteins due to the fact that the balance of their amino acids is nearer to human amino acid requirements than those of vegetable proteins. For this reason, animal proteins were formerly classified as 'first class' and vegetable proteins as 'second class' by many nutritionists.

This is a convenient but inaccurate terminology. Poor growth response in an experimental situation is due to relatively low levels of one or more of the essential amino acids in the protein under test. These are known as limiting amino acids. Suppose, as often happens in practice, that another vegetable protein rich in the amino acids lacking in the first, although poor in those abundant in the first, is taken in admixture with the first, the mixture will be utilised much more effectively than either taken separately. This phenomenon is known as a supplementary effect. Mixtures of vegetable proteins can be devised with little difficulty which are as good as animal proteins. Thus to classify animal and vegetable proteins as first and second class is misleading. If you ask a nutritionist about daily protein requirements he may not unreasonably answer by asking which protein you are talking about.

The second difficulty lies in the fact that protein intakes may vary over a period of time without obvious ill-effects. It is true that protein deficiency diseases are clinically identifiable without difficulty, but they only exist in the pooorest communities whose dietary choice is severely restricted and whose total food intake is insufficient. It is almost true to say 'take care of the calories and the proteins will take care of themselves'. Put this another way and it is certainly true to say that it is rather difficult to design a human dietary regime palatable enough to ensure that the user will meet his calorie requirement without the protein level being sufficient for his or her needs.

There have been various international attempts to establish minimum human protein requirements starting with the League of Nations Technical Committee on Nutrition which, in 1936, recommended a minimum intake of one gram of protein per kilogram of body weight per day. The most recent recommendations from the joint F.A.O./W.H.O. expert committee on nutritional standards (W.H.O. Technical Report Series no. 552, Geneva, 1973) are 0.57 g/kg for adult males and 0.52 g/kg for adult females per day, these being considered as 'safe levels of intake' calculated in terms of milk or eggs. For a standard reference man or woman these correspond to 37 g and 29 g of these proteins per person per day. Thus for a normal mixed diet giving a protein score of, say, 70 per cent nett protein utilisation, the actual requirements would be $37 \times 100/70$ and $29 \times 100/70$ or 53 g and 41 g per person per day.

1.6 The fat-soluble vitamins, A and D

Vitamins are substances required in small amounts for the proper functioning of the human or animal body which the body cannot produce in sufficient quantity.

The nomenclature of the vitamins is confused. Early workers tended to describe the characteristics of their impure preparations by designating their different physiological properties by the letters of the alphabet. This came unstuck because in some cases it later transpired that a whole group of chemical substances of similar but not identical structures could act physiologically in much the same way. In others, what was originally thought to be single vitamin from the crude preparations later proved to be a group of vitamins with separate and distinct properties. This confusion was not helped by the later discovery that some substances which were vitamins (within the above definition) for the experimental animals used, have not yet been demonstrated to be essential to man.

Yet habit dies hard, and although enough is now known to attribute chemical structures to all the vitamins important to man, the old nomenclature lingers on and trivial names persist simply because they are convenient in practice.

The systematic study of the vitamins includes some of the great and (dare I use the word) romantic discoveries of science and especially of structural organic chemistry, much of it prior to the advent of the powerful techniques of chemical and structural analysis now at the disposal of chemists and biochemists. A major problem of the early chemical workers was the fact that for all vitamins, human daily requirements were miniscule and for some (to invent a word) microscule. For example one gram of the following vitamins will meet the needs of a healthy adult for

Vitamin A	3.6 years
Vitamin D	1,100 years
Vitamin C	33 days
Vitamin B_1	2.3 years
Nicotinic acid	56 days
Riboflavin	1.6 years
Vitamin B_{12}	900 years
Folic acid	3 years

These figures are, of course, approximate. Nevertheless they reflect the problems of first identifying a given disease syndrome as being due to a vitamin deficiency and secondly of isolating and chemically identifying substances whose presence in food is at levels which are not altogether unrelated to those levels of requirement. After all the human race (or something closely akin to it) managed to survive for the two million years or so of its existence without anything better to guide it in food selection than experience coupled with its sense of smell and taste. True the race survived, but untold members died before their time from deficiency diseases which, with our modern knowledge of vitamins, are easily preventable and, when they occur, equally easily cured providing they are caught at an early stage.

I am writing this in a suburb of the city of Glasgow and I am in my middle fifties. I can walk through the shopping centre of the city and guarantee to spot (and usually within a few minutes) middle-aged and elderly men and women showing the twisted lower limbs of healed rickets of childhood. This disease was rife in my youth. It is a wholly preventable condition, due to deficiency of vitamin D, which disfigures and partially disables a person for life. Yet even now the occasional case is seen in our city hospitals.

The vitamins are conveniently grouped as water-soluble and fat-soluble. The fat-soluble group comprises vitamins A, D, E and K, while the water-solubles include the eight vitamins of the B group and vitamin C. It is still convenient to talk of the 'vitamin B group' or 'vitamin B complex' because, while its members are both chemically and physiologically distinct, they tend to occur together in the same sorts of food.

Vitamins present in food may not necessarily be absorbed in their passage through the gut. For example, much of the nicotinic acid in cereals is chemically bound and not available and fat-soluble vitamins may be poorly absorbed in cases of generally impaired fat digestion. There have been occasional reports of the presence of 'anti-vitamins' in foods. These appear to act either as competitive inhibitors of enzymes, because of their structural similarity to vitamins involved in the corresponding enzyme reactions, or by destroying vitamins in the gut before absorption. Fortunately they appear to be of rare occurrence in human diet although rather more frequent in the diets of farm animals.

The normal bacterial flora of the gut is capable of synthesising small amounts of some B group vitamins and vitamin K, but it is doubtful whether this is of much significance in normally healthy individuals. For these reasons it is not always completely accurate to define the vitamin status of a diet in terms of laboratory vitamin analysis, although in general this is a good guide and is widely used.

Vitamin A or retinol is an almost colourless compound, insoluble in water but soluble in fats and fat solvents; it is fairly stable over the times and temperatures of normal cooking but is readily destroyed in a coupled reaction if the fats which contain it are affected by oxidative rancidity. However, if vitamin E is simultaneously present it affords a fair degree of protection to this reaction.

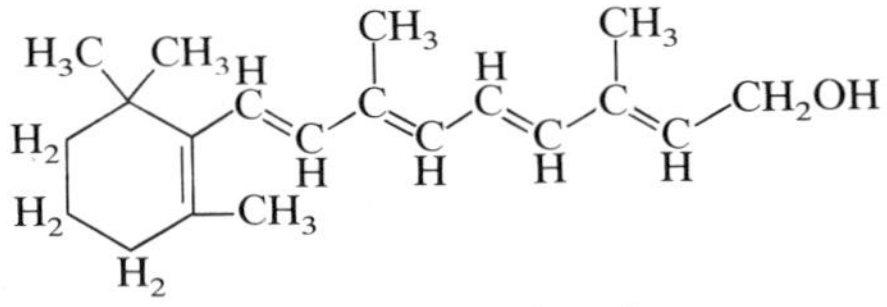

Retinol (vitamin A_1)

The carotenoid pigments, which are of widespread occurrence in plants, are structurally related to retinol and some can be wholly or partially converted to vitamin A by the body. In mammals vitamin A activity is exhibited by α-, β- and γ-carotenes and by a few other carotenoids such as cryptoxanthin. Other carotenoids such as lycopene (the red pigment of tomatoes) and xanthophyll, neither of whose ionone rings are identical with those of retinol, are devoid of vitamin A activity. Of the carotenoids, β-carotene is nutritionally the most important, one molecule being convertible to two molecules of vitamin A.

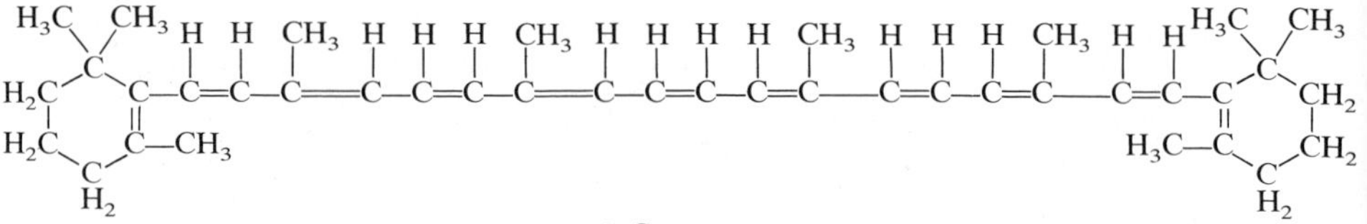

```
H3C  CH3 H  H  CH3 H  H  H  CH3 H  H  H  H  CH3 H  H  H  CH3 H  H H3C  CH3
   \ /   |  |  |   |  |  |  |   |  |  |  |  |   |  |  |  |   |  |    \ /
    C    |  |  |   |  |  |  |   |  |  |  |  |   |  |  |  |   |  |     C
H2C   C—C=C—C===C—C=C—C===C—C=C—C=C———C=C—C=C———C=C—C       CH2
 |    ||                                                  ||      |
H2C   C—CH3                                          H3C—C       CH2
   \ /                                                   \ /
    C                                                     C
    H2                                                    H2
```

β-Carotene

In α- and γ-carotene and also in cryptoxanthin only one ionone ring is identical to that of retinol and thus they only yield one molecule of vitamin A. On a weight basis they provide only half the activity of the same weight of β-carotene. However, the chemical equivalence is modified in practice by the fact that carotenoids are rather inefficiently absorbed from the gut and further losses may occur during conversion to retinol, most of which takes place in the wall of the small intestine. Thus it is probable that in effectiveness, 1 mg of retinol provides six times as much effective vitamin activity as 1 mg of β-carotene. It is still not uncommon to find vitamin A activity expressed in international units (IU), but following a joint W.H.O./F.A.O. report of 1965 it has become practice to express vitamin A activity in terms of the more precise retinol equivalents, 1 I.U. of activity being equated to 0.3 μg of retinol.

Vitamin A deficiency first appears in the form of 'night blindness' or the inability of the eye to function at low levels of illumination. In local terms the retinal pigment rhodopsin (visual purple) upon which night vision depends disappears, but this is merely the forerunner for other degenerative changes involving the skin, skeletal tissue and kidney function. Deficiency over prolonged periods leads to degenerative change in the bronchial epithelium, resulting in diminished resistance to infection. The patient thus affected usually succumbs to pneumonia. The biochemical role of retinol in preventing those changes is not clear, but the vitamin appears to play a part in the synthesis of structural protein.

The principal dietary sources are milk, butter, cheese, egg-yolk and liver. The richest natural sources of vitamin A are fish liver oils. Carotenes are also widely distributed in plants being found chiefly associated with chlorophyll in green vegetables. Of the root vegetables, carrots are a good source. Perhaps a little surprisingly, vitamin A is normally absent from vegetable oils, red palm oil being an exception.

Retinol, being insoluble in water, is not readily excreted, and if over-enthusiasm for vitamin therapy leads to an excessive intake it can produce toxic effects such as drowsiness, headaches, skin peeling and sometimes liver enlargement. Recovery is rapid following withdrawal of the vitamin.

There is a sense in which **vitamin D** is not a vitamin at all. Irradiation of the skin by sunlight or ultraviolet light has long been recognised as a successful method of treatment of its deficiency, characterised by the disease syndrome known as rickets. Early work was confused by the fact that the same disease could be successfully treated by administration of cod-liver oil, a folk remedy of long standing and used for well over a hundred years. It is now known that irradiation of 7-dehydrocholesterol in the skin produces the vitamin-D-active substance cholecalciferol, one of a number of distinct but closely related substances possessing rickets-preventing or antirachitic properties. Some of these are also present in cod-liver oil. Two antirachitic compounds are characterised below.

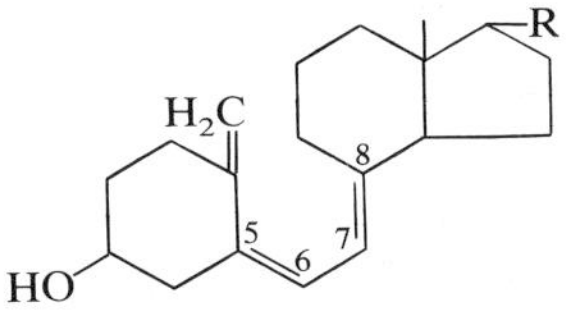

Vitamin D

$R = C_8H_{17}$ (cholecalciferol) in vitamin D_3 (from ultraviolet irradiation of 7-dehydrocholesterol)

$R = C_9H_{17}$ (calciferol) in vitamin D_2 (from ultraviolet irradiation of ergosterol.)

The term 'vitamin D' thus covers this and other variants of the same basic structure possessing antirachitic properties. The only rich sources of this vitamin are the liver oils of fish although it is also found in egg-yolk, milk, butter, vitaminised margarine and cheese.

The physiological function of the vitamin is associated with the formation of bone, and deficiency alters the processes involved in bone growth giving reduced calcification which leads to softening of skeletal tissue. Its most noted symptom is bending of the bones of the legs, which become unable to support the weight of the growing child's body, leading in many cases to life-long physical deformity. In the areas of abundant sunlight a dietary supply is not essential unless the skin is overprotected. In northern countries and even in the absence of sunlight 10 μg per day is sufficient to protect a child against rickets and adult requirements are much less, 2.5 μg being sufficient under all normal circumstances.

Cholecalciferol is now known to be the precursor of 1,25-dihydroxycholecalciferol, which is the metabolic form most active in transferring calcium across the intestinal wall. The final stage in the transformation is carried out in the kidney. Thus chronic renal failure leads to lowering of intestinal calcium absorption and the development of osteodystrophy. Knowledge of this mechanism is relatively recent and encourages hope of a therapeutically useful replacement.

If taken in excessive amounts vitamin D accumulates in tissues (being fat-soluble it is not readily excreted) and produces toxic effects which can be severe, and fatal cases have been reported. This situation generally arises from misguided enthusiasm for vitamin therapy through self-medication and is sometimes observed in children given the pure vitamin by parents accustomed to dosages associated with the use of cod-liver oil.

Vitamin D promotes the absorption of calcium from the gut but it also acts elsewhere. In the kidneys it promotes tubular absorption of phosphate. It also has the overall effect (by a controlled indirect process) of raising plasma calcium levels which in turn promote bone deposition.

1.7. The fat-soluble vitamins, E and K

Vitamin E was discovered in the early nineteen twenties by Canadian nutritionists who found that, in certain types of diets, both male and female rats became sterile and that this could be corrected by the administration of certain vegetable oils. It was not until 1936 that the pure vitamin was isolated from the unsaponifiable fraction of wheat-germ oil by the same workers. They called the substance tocopherol. The chemical synthesis of the vitamin was accomplished two years later both in Switzerland and in the United States.

α-Tocopherol

It is now known that there are several tocopherols differing from each other by the position and the number of the methyl groups attached to the ring of the molecule. All exert vitamin E activity, α-tocopherol being the most active. They are yellow oily liquids which are very stable to heat but which can be oxidised gradually in the presence of oxidising fats or oils. However, they do have the remarkable property of reducing the rate of oxidation of oils and fats in which they are present, and are therefore of interest to the food technologists as antioxidants. Since carotene and vitamin A are carried in the oily fractions of food and since these are readily oxidised in the presence of oxidising fats, the antioxidant properties of tocopherols serve the double function of inhibiting rancidity and protecting vitamin A, a matter of practical consequence to nutritionists and food technologists alike.

Vitamin E is present in human tissues and is so widely distributed in foodstuffs—even in the cheapest kind—that dietary deficiency is unlikely. Its precise physiological role is unknown although it may well be associated with its antioxidant properties. It has been fashionably prescribed for a variety of ailments ranging from heart disease to diabetes.

Athletes have been dosed with it by their coaches. It has been recommended as a panacea against some of the effects of old age. It has been recommended in the treatment of sterility and habitual abortion. There is no clear scientific evidence showing it to be of advantage for any of these purposes. Fortunately, however, there is no evidence that the vitamin possesses toxic properties, so there seems little harm associated with unorthodox use.

Vitamin K is a fat-soluble napthoquinone found in two chemical forms in nature but certain other synthesised derivatives also have vitamin activity. The natural vitamin is found in dark-green vegetables such as kale, spinach, nettles and lucerne. Cauliflower is also a good source. In general, animal products contain little of the vitamin. Deficiency of the vitamin results in damage to the normal mechanism of blood clotting with uncontrolled internal or external bleeding following the slightest injury. Primary deficiency in adults has never been demonstrated unequivocally, but deficiency may occur as a secondary consequence of other conditions such as bilary obstruction or conditions involving malabsorption such as sprue or idiopathic steatorrhea. In such cases vitamin K therapy may be required. It has also been used in the treatment of bleeding in newborn infants.

The naturally occurring molecules are substituted derivatives of menadione (2-methyl-1,4-naphthoquinone) which also is physiologically equivalent to vitamin K.

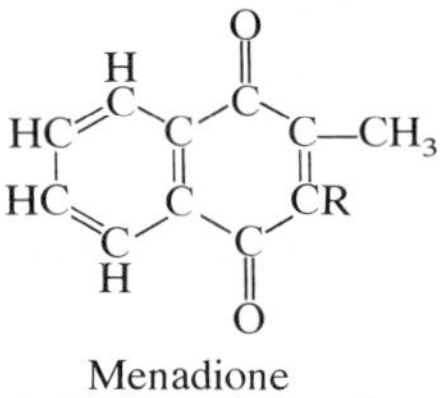

```
              O
       H      ‖
       C      C
   HC⁄   ╲C⁄    ╲C—CH3
    |     ‖      ‖
   HC     C      CR
     ╲C⁄   ╲C⁄
      H      ‖
             O
```

Menadione
(2-methyl-1,4-naphthoquinone)

Menadione R = H
Vitamin K_1 R = —CH_2—CH=C(CH_3)—CH_2[CH_2—CH_2—CH(CH_3)—CH_2]$_3$H
Vitamin K_2 R = —[CH_2—CH=C(CH_3)—CH_2]$_6$H

1.8. The water-soluble vitamins

Ascorbic acid or vitamin C was first isolated in 1928 by Szent-Györgyi from oranges, cabbage and suprarenal glands, but he did not recognise its properties as a vitamin. In 1932 Glen King isolated the vitamin from lemon juice and showed it to be identical with Szent-Györgyi's acid. Within months Haworth and Hirst in Birmingham had demonstrated its chemical structure and had found a route of synthesis. This ended more than four hundred years of trial and error in attempting to understand the cause of scurvy, that scourge of sailors. Systematic research on the disease goes back to 1753 when James Lind published the careful clinical experiments which were to demonstrate the relationship between diet and this disease. In his famous *Treatise on Scurvy* he noted that the beneficial effects of citrus fruits had been known for at least two hundred years before his time.

Scurvy, the seaman's disease, was the major hazard facing travellers during the great voyages of exploration which stretched over some four hundred years of world navigation. Scurvy was a ghastly and often fatal illness. The sixteenth century sailor Sir Richard Hawkins wrote of its identification 'by the swelling of the gums, by denting of the flesh of the legs with a man's fingers, the pit remaining without filling up in a good space; others show it by their laziness'. Later the hair follicles become enlarged and ooze blood, the teeth become loose, the gums bleed and the process of eating becomes painful in the extreme. Yet even in Hawkins' time it was known that recovery was rapid when port was reached and the men had access to green plants and fruits. As early as 1601 it was recorded that the fleet of the English East India Company hove to off the southern tip of Madagascar and gathered 'oranges and lemons of which we made good store of water (i.e. extracted juice) which is the best remedy against scurvy.'

Despite this knowledge, the carrying of expensive and, in those days, perishable fruit juice was regarded either as extravagant or impracticable by administrators and shipowners. On long voyages the mortality rates from scurvy were frightful. In 1498 the great Portuguese navigator, Vasco da Gama, reached India after a voyage out of Lisbon of almost eleven months. During the voyage, one hundred of his crew of one hundred and sixty died of scurvy. In 1577 a Spanish ship was found adrift in the Sargasso Sea with all her crew dead of scurvy. Many records of long voyages tell the same or similar stories. In 1740 the British Admiral Anson set out for the island of Juan Fernandez with six ships and nine hundred and sixty-one men. Only three hundred and thirty-five survived the journey. In the British Navy scurvy killed far more men than enemy action.

By the end of the eighteenth century the British Admiralty had decreed that a fixed amount of lemon juice be issued daily to sailors in its ships after their fifth week afloat. Thereafter the mortality rate declined with startling suddenness. Yet despite the experiments of Lind and the practical experience of the Navy, Captain Scott, R.N., planned his ill-fated expedition of 1912 to the South Pole without a single source of ascorbic acid in his rations. Retrospectively, by 1933 this must have appeared to be a stupid mistake. Yet as recently as 1963 the present writer was asked to comment on yet another expedition's rations. They were excellent in almost all respects except for the complete omission of vitamin C.

Vitamin C is a simple sugar which is identical with ascorbic acid in either its reduced or oxidised (dehydroascorbic acid) form. It is the most powerful reducing agent found in living tissue being itself oxidised in the process as follows:

```
O=C─┐            O=C─┐             COOH
  |  |              |  |               |
HOC  |             O=C  |              O=C
  ‖  O   −2H          |  O    OH⁻        |
HOC  |   ⇌        O=C  |   ───→       O=C
  |  |   +2H          |  |               |
 HC──┘              HC──┘             HCOH
  |                  |                  |
HOCH               HOCH               HOCH
  |                  |                  |
 CH2OH              CH2OH              CH2OH
```

L-Ascorbic acid L-Dehydroascorbic acid L-Diketogulonic acid

The first stage of the reaction is readily reversible but the further oxidation to diketogulonic acid cannot be reversed.

Ascorbic acid is a dietary essential for the primates (including man), for the guinea-pig and an Indian fruit-eating bat, which are the only mammals known to require it. Amongst birds, the red-vented bulbul and some other birds of the order Passeriformes also require it. In contrast, the majority of mammals, birds and reptiles synthesise their own from glucuronic acid. Creatures unable to carry out this synthesis appear to lack the enzyme to convert ketogulonolactone to ascorbic acid. There is some fairly recent evidence that some women may have a limited capacity for synthesising ascorbic acid.

```
O=C─┐              O=C─┐
  |  |                |  |
HOCH |               HOC  |
  |  O                |  O
O=C  |     ───→      HOC  |
  |  |                |  |
 HC──┘                HC──┘
  |                   |
HOCH                 HOCH
  |                   |
 CH2OH                CH2OH
```

3-Keto-L-gulonolactone L-Ascorbic acid

Experiments with human volunteers have established that small doses of ascorbic acid of around 10 mg per day not only prevent scurvy but rapidly cure the clinical features of deficiency. However, the mere suppression of the symptoms of a deficiency disease is no measure of whether the body is receiving the optimum amount for health. This optimum level is surprisingly difficult to determine. At the other extreme, there is no sustained evidence of toxicity at very high daily intakes. Human beings are reported to have taken as much as 40 g per day for a month and as much as 100 g in one day without toxic effect. In a few cases doses of several grams may cause mild diarrhoea if eaten without food but more serious side effects have not been confirmed. Between the limits of 10 mg and (say) 10 g per day there is presumably an optimum level of this vitamin, and this may vary from person to person. Various authorities have arrived at recommended daily intakes as follows for a 70-kg man.

League of Nations Technical Commission (1938)	30 mg
U.S. Food and Nutrition Board (1943)	75 mg
U.S. Food and Nutrition Board (1968)	60 mg
U.K. Department of Health and Social Security (1969)	30 mg
U.S. Food and Drug Administration (1973)	60 mg
U.S. Food and Nutrition Board (1974)	45 mg

In contrast to these figures, which represent the thought patterns of orthodox nutrition, there are some who believe that these figures may be much too low for optimum health including resistance to viral diseases such as the common cold. The controversy over this matter may not be settled for some years, since it is rather difficult to carry out closely controlled experiments on the substantial number of people required if the results are to be of statistical interest and value. Such work as has been done is indicative that there may be some justification for this hypothesis. There is at least enough evidence to justify further work along these lines.

Few physiological substances can have received more scientific attention than ascorbic acid, yet its function is still rather imperfectly understood. It is certainly required for the formation of hydroxyproline, a key amino acid in the structure of collagen. The delicate balance between the oxidised and reduced forms of the acid appears to tie in with reduced (—SH) and oxidised (—SS—) sulphydryl groups. Glutathione, a sulphydryl-containing substance, protects ascorbic acid from oxidation. Despite this, no biological oxidation system has yet been described in which ascorbic acid acts as a specific coenzyme. Its general distribution in tissues, and the fact that at saturation a human body contains about 5 g of the vitamin (such a quantity must have a number of functions one perhaps feels), and the presence of local high concentrations in the adrenal cortex, the pituitary, the liver, pancreas and spleen, all go to suggest an active role in metabolism, as yet to be accurately defined.

Thiamin or vitamin B_1 is a white crystalline substance, readily soluble in water and rapidly destroyed by heat in neutral or alkaline solutions. In acid solutions it is comparatively stable. Fortunately most foodstuffs have a pH below 7 and careful cooking avoids serious losses of this vitamin.

However, it is broken down by the action of sulphur dioxide and sulphites, which are used widely in foods as preservatives and as inhibitors of the Maillard reaction. Thiamin has the following structure.

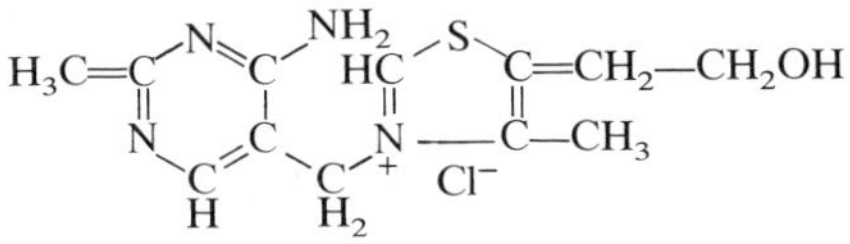

```
           N      NH2    S
H3C=C   ⁄   ⁄C  ⁄  HC  ⁄   ⁄C=CH2—CH2OH
    ||      |       ||      ||
    N      C       N ——— C—CH3
     \C ⁄    \C ⁄ +  Cl⁻
      H       H2
```

Thiamin chloride

Ruminants apart, all animals require a dietary supply of thiamin. In areas of the world where polished rice is the staple foodstuff, the disease known as beri-beri has long been endemic. In the western world thiamin deficiency in acute form is seldom seen except as a secondary syndrome amongst acute alcoholics. As nutritional understanding spreads it is becoming less common in the East, and it has now almost disappeared in the large conurbations and in the more advanced eastern communities such as Japan, the Phillipines, Malaysia, Indonesia and Singapore.

The illness occurs in three forms described as wet, dry and infantile beri-beri. In all cases the onset is insidious. At first there is loss of appetite (anorexia) associated with weakness of the legs, and the patient may complain of palpitations and an exaggerated pulse, together with either anaesthesia of parts of the leg, or of other areas of persistent 'pins and needles', and of tenderness of the calf muscles to pressure. Such a condition may persist for a considerable time but the more acute forms of the illness may be precipitated by unusual exertions, trauma, a mild feverish illness or other situation resulting in metabolic stress. Oedema then develops on the legs, face and trunk and is the noteworthy symptom of 'wet' beri-beri. Palpitations and breathlessness are marked and there is marked dilation of the heart, death being commonly due to sudden heart failure. Recovery is remarkably rapid following administration of thiamin, and within a few hours the breathing becomes easier, the pulse rate drops and rapid diuresis begins to dispose of the oedema. Within a few days the heart size returns to normal and muscular pain and discomfort are greatly improved.

In 'dry' beri-beri the characteristic feature is a polyneuropathy, the muscles becoming progressively weaker and the thin, emaciated patient eventually becomes bed-ridden. At this stage he or she is very susceptible to secondary infections, which are likely to be fatal.

Infantile beri-beri occurs in breast-fed infants of mothers on low thiamin diets and has been a major cause of death in infants of from two to five months in rice-eating rural areas in the past. The simplest method of treatment is to ensure that the mother has an adequate supply of thiamin. In such cases the mother's thiamin levels are quickly restored by twice daily injections of 10 mg of thiamin hydrochloride.

Metabolically, thiamin pyrophosphate is the coenzyme of carboxylase which decarboxylates pyruvic acid, the pyruvic acid being produced from the catabolism of glucose to produce muscular energy and heat. It is not, therefore, surprising that thiamin requirements are relatively high in high-carbohydrate diets associated with high energy expenditures. All plant and animal tissues contain thiamin and it is therefore present in all unprocessed raw foods. Yet there are only a few really rich sources, the germ of cereals, pulses and nuts, and also yeast being about the best. The vitamin is completely absent in refined sugar and in distilled alcoholic drinks, and subjects who acquire a substantial fraction of their energy needs from these sources may expose themselves to inadequate intakes of the vitamin.

With the direct involvement of this vitamin in energy release processes in the body, it is advantageous to calculate requirements in terms of energy expenditure. Thus the Joint W.H.O./F.A.O. Committee (1967) recommended a daily intake of 0.4 mg/1,000 kcal (4,000 kJ) of food energy consumed. American recommendations tend to be a little higher at, for example, 1.4 mg per day for an average 70-kg man aged between 23 and 50 years.

Once upon a time there was a **vitamin B_2** preparation. In 1933 Khun and his colleagues isolated the pure crystalline substance **riboflavin**, 1 g being obtained from around 5,500 l of whey. However, this substance had some but not all of the properties of the original vitamin B_2 preparations, which on further study were found to contain other vitamins in addition to riboflavin. The term 'vitamin B_2' is still sometimes used to describe riboflavin but is perhaps better avoided.

Riboflavin is a yellow crystalline substance, insoluble in fat but slightly soluble in water. It is readily decomposed by heat in alkaline solutions but is stable at boiling point under acid conditions. It can be destroyed by exposure to light. The best dietary sources are liver and kidney, yeast, cheese, eggs and green vegetables, but cereals and milk also contribute usefully. It is fairly widespread in its occurrence in foodstuffs and acute illness in man from uncomplicated deficiency of this vitamin is almost unknown. Its structure is as follows

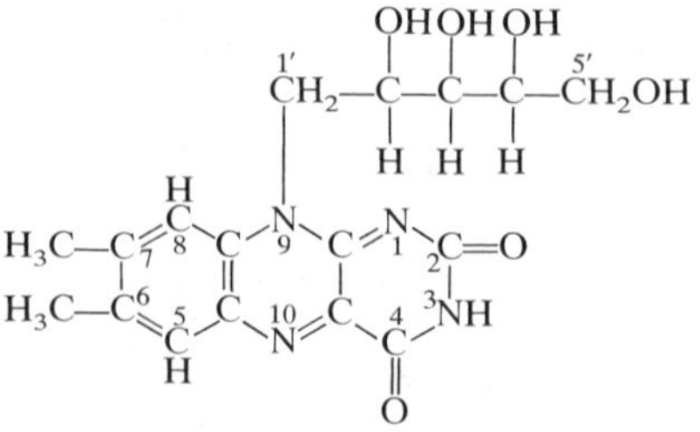

```
                     OH OH OH
                 1'  |  |  |     5'
                 CH2—C——C——C——CH2OH
                  |  |  |  |
                  |  H  H  H
          H       |
          C       N       N
H3C—C7  8   C   9   C   1   C=O
    |       ||      |      2|
H3C—C6  5   C   10  C   4   NH
          C       N       C
          H               ||
                          O
```

Riboflavin
6,7-dimethyl-9-(1′-D-ribityl)isoalloxazine

The flavoproteins are stable combinations of riboflavin with a number of different proteins and are coenzymes essential for the oxidation of carbohydrates. They act as intermediate hydrogen carriers in the process and include cytochrome *c* reductase, flavin mononucleotide and flavin-adenine dinucleotide.

Deficiency in humans results in skin cracking at the corners of the mouth and on the lips, and degenerative changes of the tongue associated often with similar changes in the neighbourhood of the anus and genitals.

Recommended intakes are 0.55 mg/1,000 kcal (4,000 kJ) or about 1.5 mg per day.

Nicotinic acid (formerly known as niacin) and its related substance **nicotinamide** have the same biological function and the term niacin is now applied to both of these substances. They function as a coenzyme in the metabolic oxidation of carbohydrates in the form of nicotinamide-adenine dinucleotide. The acid is a white crystalline substance, readily soluble in water and resistant to acids, alkalis and heat. It has the following structure

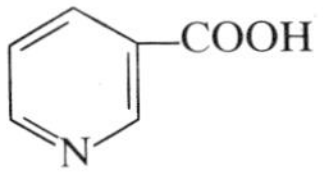

Nicotinic acid

and can be formed in the body from tryptophan, 60 mg of tryptophan yielding 1 mg of niacin. It is found in yeast, meat and liver and also in cereals, although much of that present in cereals may be in a bound form and unavailable to human digestive systems. Bound nicotinic acid in cereals is released by alkali treatment, a fact of some significance in Mexico and Central America where maize is usually treated with lime before baking into tortillas.

Niacin deficiency is the principal cause of the disease pellagra, although other deficiencies may also contribute to the syndrome. In the early part of the present century pellagra was common in the maize belt of the United States, an average of 170,000 cases being reported annually. The symptoms include dermatitis, diarrhoea and mental disturbance.

Dietary requirements for niacin are placed at 6.6 mg of nicotinic acid/1,000 kcal (4,000 kJ) in the diet, corresponding to about 18 mg per day for an adult male.

Pyridoxine and its related compounds **pyridoxal** and **pyridoxamine** are collectively known as **vitamin B_6**. Physiologically these serve as coenzymes or prosthetic groups for aminotransferase enzyme systems. In deficient diets, experimental animals show symptoms of a degenerative nature which vary (depending on the kind of animal) from dermatitis and anaemia to peripheral neuropathy and epileptic convulsions. In man deficiency is rare, being occasionally observed in infants. There are also reports of ailments such as dermatitis, general weakness and insomnia responding to administration with pyridoxine, but these reports usually refer to patients on poor diets and there are no characteristic symptoms of deficiency known in adults.

Human requirements for vitamin B_6 are not known with precision but are thought to be of the order of 2 mg per day from the observation that most diets of healthy adults provide about that amount.

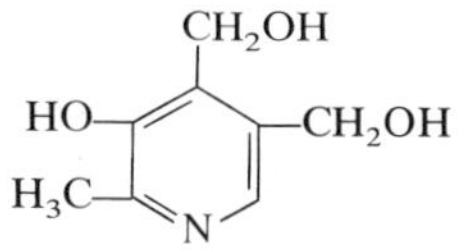

CH_2OH
HO— —CH_2OH
H_3C— N

Pyridoxine

Pantothenic acid is present in all living matter and its wide distribution in natural foodstuffs ensures that deficiency is unlikely to arise except in abnormal circumstances. There is no clear description of deficiency of this vitamin in humans although induced deficiency in experimental animals produces a range of symptoms from dermatitis and greying of hair to neuropathies and myelin degeneration. Pantothenic acid forms part of the structure of coenzyme A and is therefore required for the metabolism of carbohydrates and fats. Human requirements appear to be of the order of 6 mg per day.

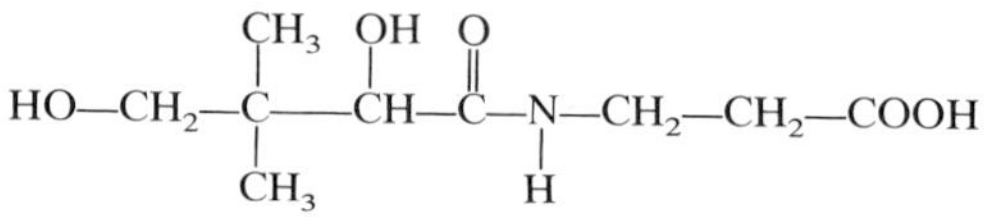

$$HO{-}CH_2{-}C(CH_3)_2{-}CH(OH){-}C({=}O){-}NH{-}CH_2{-}CH_2{-}COOH$$

Pantothenic acid-(pantoyl-β-alanine)

Biotin is of biochemical interest as the prosthetic group of a series of carboxylase enzymes but is of little practical concern to nutritionists since human deficiency is virtually unknown. It seems likely that man can obtain his requirement from the activities of his intestinal bacterial flora. However, biotin does combine with avidin (a protein of raw egg-white) and is unavailable in this form. A case has been reported of a man who lived on six dozen raw eggs weekly washed down by four quarts of red wine daily. He developed a severe dermatitis which responded to biotin injections.

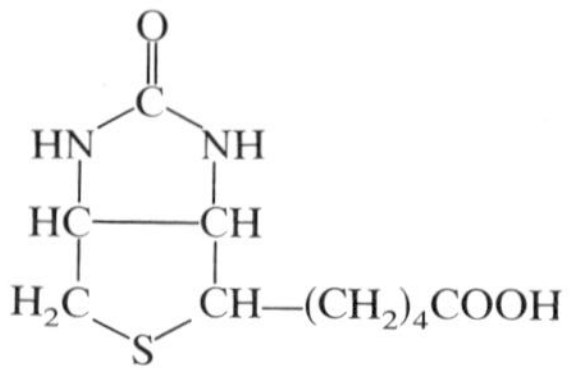

O
C
HN NH
HC—CH
H_2C CH—$(CH_2)_4COOH$
S

Biotin

The two remaining B group vitamins, **folic acid** and **cyanobalamin** are more significant in practical nutritional terms than the previous two and deficiency of either or both is associated with nutritional anaemias.

Folic acid (also referred to as folacin) is an essential vitamin in the synthesis of purines, pyrimidines and some amino acids.

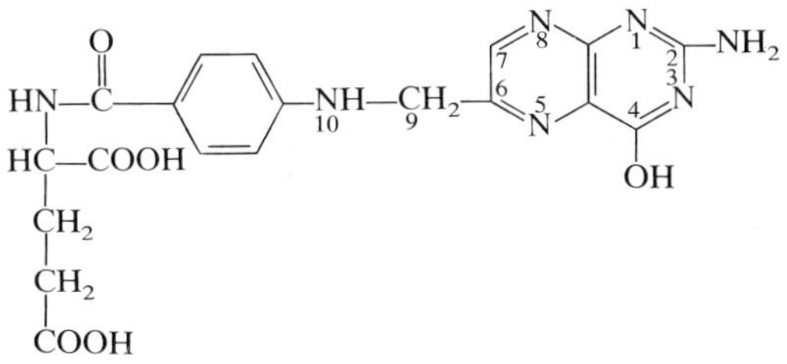

Folic acid (pteroylglutamic acid)

It occurs in foods as a variety of derivatives, some of which have been given separate and rather confusing trivial names. Activity is found in liver and kidney and in green vegetables, and the assay is subject to uncertainties due to the variety of chemical forms in which the vitamin occurs. The term folacin is used generically for dietary sources as assayed by *Lactobacillus casei* and the requirement for adult males is estimated (Food and Nutrition Board, U.S. National Academy of Science, 1974) as 0.4 mg per day. However, when administered in the form of pteroylglutamic acid one-quarter of this level is believed to be effective.

The megaloblastic (abnormal and enlarged red blood cells) anaemias associated with folic acid deficiency were first observed in pregnant women in India, living on a rice and wheat diet. These responded to treatment with autolysed yeast but not to purified liver extracts (which would be a good source of vitamin B_{12} and were used at one time in the treatment of pernicious anaemia). Megaloblastic anaemia due to simple dietary deficiency of folic acid is rarely seen in wealthy countries although it occasionally arises during pregnancy when the requirement appears to be increased. The disease is common amongst deprived peoples of both sexes in the tropics, although women and children are more often affected. In Britain it is occasionally encountered amongst the housebound elderly, living alone and on poor diets.

Vitamin B_{12} or **cyanocobalamin** is absent from higher plants and found only in animal foods, certain fungi and some seaweeds. Each molecule contains one cobalt atom. Dietary deficiency is rarely observed except in vegans (i.e. vegetarians who strictly avoid all foods of animal origin). Recommended dietary allowances are 3 μg. Pernicious anaemia is due to malabsorption of the vitamin due to failure of the gastric mucosa to secrete as yet unidentified mucoproteins known collectively as the 'intrinsic factor'. This is required for the transport of the vitamin across the intestinal wall. Pernicious anaemia, one-time a fatal illness, is now readily controlled by monthly injections of the purified vitamin. Cyanocobalamin is the most complex in structure of the vitamins and its elucidation by Hodgkin at Oxford and Todd at Cambridge is one of the triumphs of modern organic chemistry.

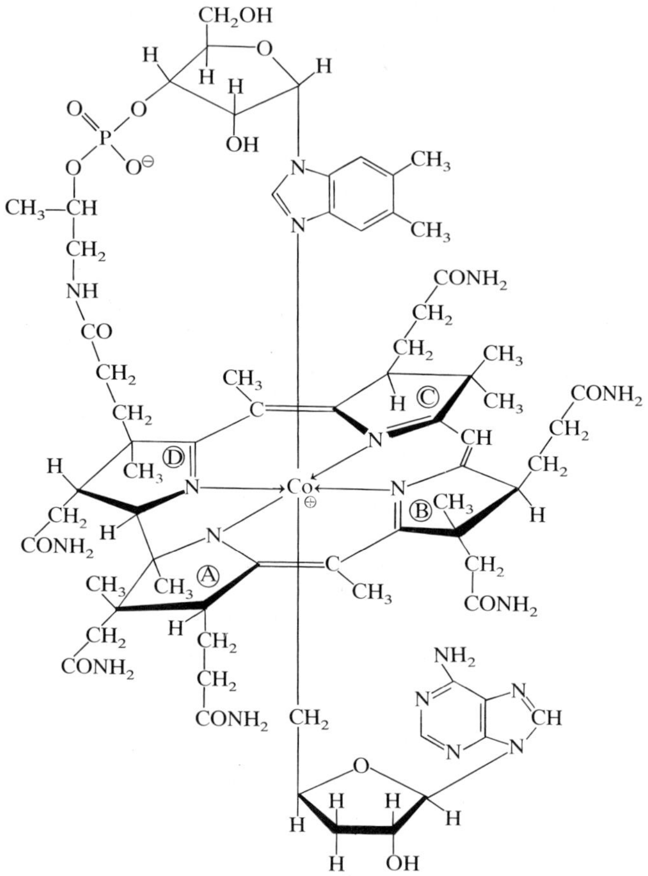

Cobamide, the coenzyme form of vitamin B_{12}

Despite the widespread scientific interest in folic acid and cyanocobalamin, simple iron deficiency is by far the most common cause of anaemia in most temperate parts of the world. In uncomplicated cases, treatment by 200-mg tablets of ferrous sulphate taken three times per day is cheap and effective.

Treatment of other forms of anaemia is also normally simple and effective, once an accurate diagnosis has been made. However, this is not always quite such a simple matter.

Further reference

Manual of Nutrition (1976) 8th edn, Her Majesty's Stationery Office, London.

McCance and Widdowson's The Composition of Foods (1978) Paul, A. A. and Southgate, D. A. T., 4th edn, Her Majesty's Stationery Office, London.

Cooper's Nutrition, Health and Disease (1968) Mitchell, H. S., Rynbergen, H. J., Anderson, L. and Dibble, M. V., 15th edn, Pitman Medical, London, and Lippencott, Philadelphia.

Hutchison's Food and the Principles of Nutrition (1969) Sinclair, H. M. and Hollingsworth, D. F., 12th edn Edward Arnold, London.

Nutritional Problems in a Changing World (1973) Hollingsworth, D. F. and Russell, M. (eds), Applied Science, London.

Human Nutrition (1976). Burton, B. T., 3rd edn, H. J. Heinz Company and McGraw-Hill, New York.

Human Nutrition and Dietetics (1975) Davidson, S., Passmore, R., Brock, J. F. and Truswell, A. S., 6th edn., Churchill-Livingstone, London.

Dictionary of Nutrition and Food Technology (1975) Bender, A. E., 4th edn, Newnes-Butterworths, London

2

The Cereals

2.1. Introduction and cereal grain structure

Organised farm settlements began in the Middle East about ten thousand years ago—a mere yesterday in the million year history of humankind. But the crops of wild wheat and barley that are still to be found in certain parts of this area were doubtless used long before deliberate cultivation began. In recent years excavations of ancient settlements in the area have unearthed scorched grains which have survived the passage of time. Furthermore dwellings were built of mud bricks reinforced with straw. Occasional grains remained attached. Although grain and straw have long since crumbled to dust, the cavities they left carry imprints almost as detailed and informative as the grains themselves. From these and from the evidence of implements such as flint or obsidian-edged sickles and primitive grinding stones, we can build up a picture of the early uses of cereals. From it a case can be made for suggesting that civilisation itself depended on the use of cereals from its earliest beginnings. From the hundred square miles required to provide food for the hunter-gatherer to the few acres needed by the early farmers to feed the same number of people was the step that made possible high-density living. The subsequent division of labour allowed for the development of new craft skills, of social structure and organisation and later of the arts including the important art of writing. None of this could have happened without the adaptation of cereal crops to human need.

Ten thousand years of trial and error, and a hundred years of scientific breeding have gone to the making of the modern cereals. The varieties are therefore numerous to the point of bewilderment. For example over thirty thousand strains and varieties of the genus *Triticum* (wheat) are said to exist. The other principal cereals are rice, maize, barley, oats, rye, sorghum and millet. All are cultivated forms of the great family of the grasses, that monocotyledonous group of plants known botanically as the *Gramineae.*

The significance of cereals in human nutrition can best be quantified by examples. Thus, in 1972 the eight major cereals summed to a grand world production total of around 1150 million tonnes, which was sufficient to provide about 850 g per person per day for every man, woman and child then on the surface of the earth. But the significance of cereals in human diet varies with the economic status of the different countries concerned. In many rural areas including large parts of Africa and Asia cereals provide 70 per cent or more of the energy in the diet. However, the proportion tends to fall with rising prosperity. Typically about 25 per cent of the dietary energy will be derived directly from cereals in North America and in many European countries. However, insofar as cereals are also used in developed countries as feed for livestock, the overall dependence is greater than the direct usage figures may suggest.

The term 'corn' as applied to cereals varies in meaning depending on which cereal dominates in the area in which the word is used. For example, in England the word describes wheat, in Scotland oats and in the United States of America, maize.

Cereals are used in a wide variety of forms, and even the word 'bread' means different products in different cultures. The whole grains can be cooked in water and served as rice is commonly served today. They can be roughly crushed to a coarse meal and cooked in the form of porridge. These forms are not much favoured nowadays especially if the outer layers of bran are not removed before cooking. The flavour can be altered and enriched by 'parching' or heating the whole grain to a point at which some decomposition takes place. In some instances this improves the ease of separation of the bran as well as altering the flavour of the original cereal. Most commonly nowadays, cereals are reduced to a fine powder or flour by some form of grinding process, in which some or all of the coarser bran particles are removed by sieving processes introduced between different grinding stages.

The term 'meal' is used to describe the product obtained by milling or grinding whole cereal grains. The term 'flour' refers to the fine powder obtained after a process in which milling is followed by size-separation of coarser bran particles leaving a whiter and 'purer' flour as the principal end-product of the process.

Because of their family relationships, it is not altogether surprising to find that the different cereals are quite alike in terms of their gross chemical characteristics. Thus most samples of cereals will fall within the following general range of composition

Moisture	10–14%
Protein	7–12%
Carbohydrate	63–73%
Crude fibre	1– 4%
Fat	1– 5%
Ash	1.5–2.5%.

In addition, cereals contribute significant amounts of vitamins and minerals to human diet.

It has been estimated that between 4.5 and 5 per cent of the total land surface of the earth is used for cereals and Table 2.1 (adapted from Kent, 1975) gives a broad picture of the relative economic importance of the different crops.

Table 2.1. World cereal crops in terms of relative output, areas used and yield

Cereal	Percentage of total production	Percentage of total area	Average yield	
	%		cwt/acre	tonnes/ha
Wheat	33	36	12.0	1.5
Maize	26	18	19.5	2.4
Rice	14[a]	18	10.0[a]	1.3[a]
Barley	13	12	14.2	1.8
Oats	6	6	13.4	1.7
Sorghum	5	6	9.6	1.2
Rye	3	4	11.4	1.4

[a] Rice is harvested as paddy but rice production figures are generally given in terms of milled equivalent, which averages around 70 per cent of the paddy weight.

The wheat endosperm (see Fig. 2.1) forms about 82 per cent of the grain weight, the outer layers of bran together with the aleurone cells account for a further 15 per cent, the remaining 3 per cent being the germ tissue consisting of the embryo embedded in the surrounding scutellum. (It is an instructive experiment to excise the germ tissue of a dozen or so wheat grains from a recent crop. This can be done quite easily using a razor blade. If these are now placed on damp filter-paper in a petri dish, they will germinate in a few days, although most will subsequently die of starvation, separated from their life-support system in the endosperm). The relative proportions of bran, endosperm and germ vary from one

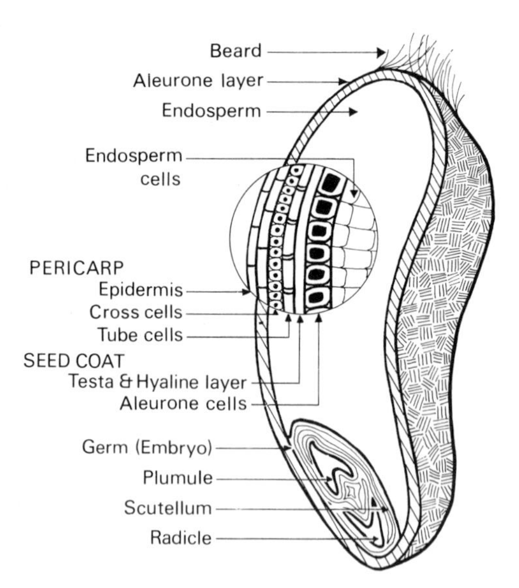

Fig. 2.1. A longitudinal section in diagram form of the wheat grain cut through the crease

cereal to another but in all cereals the endosperm is by far the largest part of the grain.

It is also the most highly prized part from the point of view of the human consumer, and in general milling processes are designed to separate the endosperm from other parts of the grain with a view to using the endosperm for human food, the remainder going for use as cattle fodder.

While the anatomical structure of all cereal grains is essentially the same, wheat, rye and maize consist of the fruit coat (pericarp) and seed which in turn consists of the seed coat (testa) germ and endosperm. This type of structure is known as a naked caryopsis. However, in the case of oats, barley and rice there is an additional cover or husk which consists of the fused glumes (or palea and lemma) surrounding the caryopsis or kernel of the grain. In naked caryopsis seeds the glumes tend to remain attached to the seed-head of the original plant during threshing, or to fall away and to be easily separated as chaff.

Each of the main parts of the grain is subdivided into identifiable anatomical portions or layers as follows.

Kernel (caryopsis)

1. The pericarp consists of an outer epidermis (epicarp) covering remnants of intermediate thin-walled cells above a layer of cross cells and a lower layer of tube cells.

2. (a) The seed proper consists of a seed coat (testa) covering a hyaline layer. Below this and immediately surrounding the endosperm is a layer of large aleurone cells. Although the endosperm itself is often described as starchy, it normally contains a substantial proportion of protein.

(b) The germ or embryo lies at one end of the grain and consists of a plumule (which on germination develops into the shoot of the plant) and radicle (which develops into the root). These can readily be distinguished microscopically as being embedded in the surrounding tissue known as the scutellum.

2.2. Wheat and wheat milling

Wheat is by far the most suitable cereal for bread making and its superiority stems from the physical properties of its proteins. To use an imperfect analogy, these have some of the elastic properties we more normally associate with rubber. Thus when carbon dioxide is produced by yeast fermentation, much of the gas is retained in the form of bubbles which expand during baking to give the fine-meshed structure prized for the excellent texture it confers on the finished loaf. To a minor extent, rye shares this property with wheat but even the best rye-breads are heavy and coarse compared to wheat. The other cereals may be used for biscuits or unleavened bread or cooked and eaten in the form of grains as is rice. None compare with wheat in bread making.

For ten thousand years wheat grains have been crushed or milled to produce the powder or flour which was the starting-point for further invention in terms of the wide range of foodstuffs prepared from it. In the earliest times the grain was either pounded to a powder in a primitive mortar and pestle or ground by rubbing a large oval water-rounded stone (such as can be found on many beaches or river beds) over the grains sprinkled over an even larger flat slab of stone. Such a process is very laborious. As skill in stone working improved it was found that the task was made easier if grooves were cut in the stone to roughen the surface. Later improvements led to the introduction of circular stones which could be pushed by slaves using capstan-bars inserted into radial holes to provide leverage. From this it was only a step to harness a mule or horse to the task of rotating the upper stone. By the time of the Roman Empire all these advances had been made. The need to provide machines to grind corn may have been the major motivation in early engineering studies.

In medieval times animal power gave way to the water-mill or windmill. Coarsely ground meals were passed through sieves of wire or cloth to remove the larger particles of bran. The resulting 'white' flours were relatively expensive and became the food of the wealthy, the coarser residues going to make bread for the peasant or labourer. In 1860 roller-milling (as opposed to stone-grinding) was introduced as a steam-powered Hungarian invention. It was a great improvement on previous processes and spread rapidly thereafter to other countries. The roller mill made white flour cheap and available to all. White breads now dominate the markets in almost all wheat-eating communities. Ironically, a reverse trend is perhaps discernible in the United Kingdom in that wholemeal bread is mostly consumed by the wealthy and fashionable!

The technology of wheat flour milling is not the concern of a text on food science but an outline account is necessary for an understanding of the significance of certain properties of wheat and wheat flour.

As received at the mill, wheat normally contains extraneous matter such as mud, stones, weed seeds, binder twine, straw, chaff, rodent excreta, insects and metal objects such as an occasional nail or nut and the like. It may also contain fungal impurities which, if ground with the flour, will spoil the taste or even, as in the case of ergot, make it dangerous to use for human food. Although these impurities make up a bizarre list they seldom amount to more than 1 or 1.5 per cent of the delivered weight of the grain, but they must be removed before milling can begin. This cleaning operation is carried out in the screenroom of the mill.

Impurities larger or smaller than wheat grains are removed by screens or sieves mounted in frames which are mechanically shaken in a reciprocating or gyratory movement. Particles, longer or shorter than wheat grains, can be removed by disc separators. The surfaces of the vertically rotating discs are provided with a series of indentations deep enough to accommodate the impurities but not the wheat grains. These revolve in the moving bed of grain, the impurities fall from the holes when they pass the top centre position into suitably placed troughs from which they are removed continuously. Other types of cleaning machines use the terminal velocity of the grains as they fall through an up-rising current of air to blow of small seeds having a terminal velocity lower than that of wheat, and by careful regulation of the air velocity good separation can be obtained. Electrostatic separators and electronic colour-sorting machines have also found application in the screenroom.

In addition to processes of physical separation, wheat may be water washed, conveyed by means of a worm-screw to a centrifuge (which the millers usually call a whizzer), agitated and then spin-dried. There is a gain of moisture during this process of around 3 per cent. Apart from removing dried caked mud and similar impurities, the moisture gain may be useful in preparing the grain for the milling process proper.

Since the miller wishes to produce flours of near-constant characteristics whether these be purpose-aimed at bread, cakes or biscuits, and since the raw material of his trade is variable, he blends wheats of different types in his grist (or mixture fed to the mill) adjusting the mixture from the range of wheat types available to him, as circumstances dictate.

The behaviour of the grist during milling is markedly influenced by its initial moisture content. As this increases the bran tends to become tougher and less brittle and the endosperm becomes more friable. However, the degree of cohesion between bran and endosperm increases so that the bran is less easily detached. In addition, too high a moisture content makes separation by sieving processes more difficult. There is thus an optimum moisture content for a given grist at which milling is accomplished easily and with clean separation of fractions at the various stages.

Furthermore, during milling moisture is lost by frictional heating and by the exposure of large surfaces to the air. This loss varies between 1 per cent and 2.5 per cent. The optimum moisture in the finished flour is around 14 per cent. At very low moisture contents (12 per cent or less) the flour is subject to the development of off-flavours due to oxidative rancidity of the fat present. At 15 per cent musty mould-like flavours develop and at higher levels active mould growth and heating and sweating may take place.

Thus the grist must be adjusted for moisture content by a process known as conditioning before being fed to the mill. In general the hard spring wheats (see page 52), used for bread flours, are milled at higher moisture contents than the soft winter wheats used for cake and biscuit flours.

After wheat has been wetted in passing through a whizzer, the outer layers may be moist while the inner layers are unaffected. Complete equilibration of the added water may require one to three days with regular turning of the grain. The movement of moisture within the grain and its equilibration is greatly hastened by gentle heating and most modern processes depend on warm conditioning of one sort or another, temperatures of around 46°C (115°F) being applied for times of 1 to 1.5 hours. After treatment the wheat is usually rested for 24 hours before milling. The process is normally carried out continuously in a vertical tower divided into sections for preheating, conditioning and cooling.

The milling process appears complex on making one's first visit to a mill but it depends on successive applications of a simple principle. In the first section of the mill the aptly named break rollers break the grain open using the shear forces developed between two closely set grooved rollers, one revolving at two and a half times the speed of the other. Both rollers carry shallow grooves set along the length of the roller. The miller usually refers to these as 'fluted' rollers and the design of the flutes affects the ease of grinding. The grinding effect is enhanced by setting the flutes at an angle to the axis. In the United Kingdom a spiral of 1 in 7 is generally used. That is to say, the flutes are cut 1 cm around the roll for every 7 cm of its length. The degree of spiral used varies from one country to another as also does the profile of the grooves. The combined effect of opposed spiral and groove is to produce a scissors-like action in the 'nip' (i.e. the gap) between the rolls, which tends to split the grain open.

The process is carried out in (normally five) progressive stages. In the second section a series of pairs of smooth rolls are used to reduce the particle size to the fine dimensions of flour. In this reduction section, shear stress is largely replaced by compression, although a speed differential between the rolls of 1.25 to 1 does retain a certain element of shear in the process.

In any grinding or size-reduction process the presence of small particles (or fines) produced in the early stages cushion the grinding effect of the mechanical action used. For efficient grinding, fines must therefore be removed as fast as they are produced. In practice, some form of size separation must take place after each grinding stage. In a typical flour milling sequence, five sets of break rolls are followed by seven or eight sets of reduction rolls. If each stage is followed by sieving into, say, three fractions, at least 12×3 or 36 different product streams will result. However, some of the streams from the 'head' of the mill will feed rolls

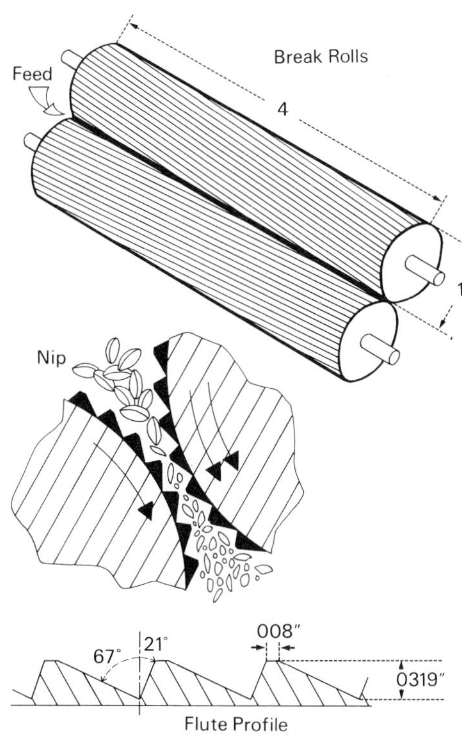

Fig. 2.2. Break Rolls showing action of flutes at the nip and a typical flute profile. Profiles and flutes become smaller further down the break system. The ratio roll-length to diameter is usually about four

further down towards the 'tail' so that the process is not quite as complex as it appears at first sight. Furthermore each set of rolls produces some flour, the amount progressively increasing as the feed stream moves down the mill, and the flour streams produced at each stage are later recombined to produce the finished product. Likewise at each stage some bran is separated and the collective result of this is to produce the millers' offals, which are mostly used in cattle feeds. However, the bulk of the product from the break rolls is endosperm of varying degrees of fineness with bran of varying size and amount still adhering to some of it. This is the material which goes for further grinding and separating further down the mill.

The effectiveness of the process depends on the fact that the bran coats are relatively fibrous and tough compared to the endosperm. The gradual progressive reductions in particle size as the material passes through the break system, combined with the shearing nature of the forces applied, tend to keep the bran particles relatively large and flat so that they can be readily separated from the more chunky particles of endosperm. After the first break much of the bran is adhering to endosperm. However, the gap between the rolls and flutes cut in them becomes smaller through the sequence. Thus at each stage more and more endosperm is scraped clear of bran. By the fifth break roll almost all the endosperm has been removed from the bran.

The milling process can thus be seen as a large-scale dissection technique for separating bran and endosperm. Furthermore it is possible to separate bran and germ to produce a fairly clean germ stream. Intermediate mixtures of streams are also obtainable. This fact has an important influence on the nutritive value of flour and bread, since the nutrients are not uniformly distributed between the different anatomical parts of the grain.

White flour is pure or almost pure endosperm. If 60 kg of flour is produced for every 100 kg of wheat fed to the mill, the miller describes this as a 60 per cent extraction rate. As the rate rises (and especially above 70 per cent) progressively more bran, aleurone and even germ is or may be incorporated in the flour. Above about 72 or 74 per cent extraction the product colour deteriorates fairly rapidly owing to the incorporation of non-endosperm constituents. Since flour fetches a higher price than offals, millers work at the highest extraction rate which will give them a flour colour their customers will accept. The actual extraction rate obtained in practice is influenced by the quality of the grist used, but it is typically in the neighbourhood of 70 per cent or just a little over.

2.3. Maturation of wheat flour

The storage of raw flour for a period of some months after milling gives rise to an improvement in its bread-baking properties. The loaves produced from matured flour are whiter in colour, bigger in volume (from the same weight of dough), and of better internal structure. Flour and bread colour is not only influenced by the presence of bran in the flour. The endosperm contains carotenoid pigments (mainly xanthophyll and its esters) which give the flour a creamy colour and impart a yellowish tinge to bread made from it. On exposure to the atmosphere these pigments are slowly bleached by oxidation, the process being even slower when the flour is stored in bulk. In consequence of these changes bread baked from flour aged or matured by simple storage has a brighter and more attractive colour.

Storage has other advantages. Changes also take place in the flour lipids and proteins. There is an initial increase in the levels of free fatty acids. At a later stage oxidation products of fatty acids appear and the proportions of linoleic and linolenic acids in the lipids decline. At the same time the number of disulphide bonds in the wheat proteins declines and the number of sulphydryl (—SH) groups increases. These changes and perhaps others alter the balance of the viscoelastic properties of the wheat proteins so that the carbon dioxide produced during fermentation is more effectively retained within the dough structure, giving increased loaf volume and improved texture.

This improving effect on baking quality (sometimes called 'aging') can be greatly speeded up by adding small quantities of chemical substances at levels normally around 10–40 parts per million of the flour weight. It is convenient to remember that 1 part per million corresponds to 1 g per metric tonne and approximates closely to 1 g per ton avoirdupois. The following are at present permitted under British regulations: ascorbic acid, potassium bromate, ammonium or potassium persulphates, chlorine dioxide, azodicarbonamide, L-cysteine hydrochloride and benzoyl peroxide.

It will be noticed that most of these substances are oxidising agents. Much effort has been given to the elucidation of their action, but their effects are not identical. In general terms it is sufficient to say that they oxidise the cysteine thiol (—SH) groups present in the wheat gluten. These oxidised groups are thus no longer available for exchange reactions with the disulphide groups (—SS—) already present in the gluten and which, by binding protein chains in the giant molecules present, contribute to its viscoelastic properties. Disulphide interchange is understood to be the mechanism of release of internal stress within the dough after mixing. If this is inhibited, the stresses are not released and the dough is 'tightened', to use the baker's descriptive term. In this state the dough's extensibility is reduced (i.e. it stretches less easily), and its elastic properties (its tendency to recover its original shape when deformed) are increased. Its ability to retain gas depends on an optimum balance between these two properties. For this reason, too high a level of chemical treatment does not have the desired effect, but merely results in the dough becoming so 'tight' as to be unworkable. Thus the use of such chemicals is self-limiting.

Ascorbic acid is a reducing agent being itself oxidised to dehydroascorbic acid which in the presence of the enzyme dehydroascorbic acid reductase is able to oxidise the sulphydryl compounds present. L-Cysteine is also a reducing agent and is used together with the slow-acting potassium bromate as oxidising agent. The L-cysteine function is to speed up the uncoiling of the protein molecules thus exposing them to the bromate effect, which stimulates the formation of the molecular three-dimensional structure, optimum for baking properties, by oxidation of —SH groups. Its main use is in the so-called 'activated dough development process', which aims to produce bread of good conventional character without the need for the costly bulk fermentation step traditionally used for this purpose.

This section started with flour colour but rapidly moved to the effects of improvers. At this point the reader may feel a little confused about the nature of the agents used to accomplish these results. Oxidising agents, such as benzoyl peroxide, are soluble only in the lipid phase of the flour or dough and act solely as bleaching agents, the pigments being lipid-soluble. Others, such as the persulphates, azodicarbonamide, potassium bromate and ascorbic acid, are water-soluble, act only on the proteins and therefore act solely as improvers without any bleaching effect. Chlorine dioxide has the advantage of solubility in both phases. It thus both bleaches and improves the flour.

Because bread is such an important item of diet in most countries, anxieties have been expressed from time to time about the safety of bleachers and improvers, despite the tiny quantities used. However, such substances have a long history of apparently safe use and no ill-effects have been attributed to human health as a result. Nitrogen trichloride (NCl_3) is a particularly useful combined bleaching and improving agent which was in common use until 1955. Its use has been discontinued as a result of evidence that, when applied at very high levels of bread, the product produced nervous symptoms in the form of hysteria in some experimental animals, the most extreme reaction being in the case of dogs. The toxic substance was shown to be methionine sulphoximine.

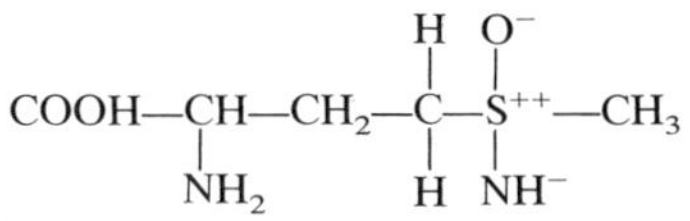

$$\mathrm{COOH{-}\underset{\displaystyle NH_2}{\underset{|}{CH}}{-}CH_2{-}\overset{\displaystyle H}{\overset{|}{\underset{\displaystyle H}{\underset{|}{C}}}}{-}\overset{\displaystyle O^-}{\overset{|}{\underset{\displaystyle NH^-}{\underset{|}{S^{++}}}}}{-}CH_3}$$

No corresponding symptoms have ever been demonstrated in man. Nevertheless it is sometimes argued that the use of all bleachers and improvers in flour should be discontinued. If this were imposed by legal restrictions the effects would be either a deterioration in bread quality or an increase in its price, due to the high cost of storage if natural aging were to be used or the high cost of equipment if physical means (such as high-speed mixing) alone were permitted. Over the years, much work has been done on achieving the same effects as chemical bleachers and improvers by means of enzymic reactions and special processing methods. There is no doubt that good bread can be so produced but only at increased costs.

2.4 The chemical composition of wheat and the nutritive value of wheat and flour

The gross chemical composition of the wheat grain varies with variety (some tending, for example, to higher protein levels than others), climate (winter-sown and spring-sown wheats tend to differ in character) and with soil nutrient status (other things equal, a given variety tends to produce a higher protein content in the grain when grown in a nitrogen-rich soil). The chemical composition of the grain will also be affected by harvesting conditions which influence levels of enzyme activity in the grain. These in turn have a major effect on the baking qualities of the resulting flour.

In dealing with any of his raw materials the food technologist is concerned with two aspects of quality. On the one hand, the nutritional value of the product is almost always dependent on that of the raw material he started with. At the same time, those who buy his product are, in the first instance at any rate, principally interested in its eating qualities. The food scientist usually refers to these as organoleptic properties and classifies them in terms of general appearance together with colour, odour, taste and texture.

It is right and proper that the ultimate user should base his judgement on these properties rather than on an academic appraisal of nutritive value. In the million-year history of mankind, appearance, smell and taste were the only criteria available to him in selecting food. That they served him well is testified to by his present dominance in the natural world. However, to the professional food scientist, concerned not with feeding one family only but with feeding nations, nutritional value becomes a matter of significance which will influence all his judgements on the use and applications of his skills. The importance of the nutritional approach is shown clearly in dealing with a staple food which represents a major part of a nation's diet. The nutrient content of flour and bread depends in part at least on how wheat is milled and flour is baked. For this reason most governments make regulations of one sort or another to control some of the factors which influence the nutritive value of essential foodstuffs for the benefit of their citizens. This is laudable if the issues are clear-cut and if all the scientific knowledge necessary for good decision making is available. In practice, the issues are seldom as clear as a superficial look suggests and the scientific evidence is often less complete than the scientists themselves would wish.

Table 2.2 is compiled from data obtained by McCance and his co-workers in a classical study of the effect of extraction rate on chemical composition of flour reported in 1945 (*Biochemical Journal*, 1945, *39*, 213). The data are calculated to a 15 per cent moisture basis.

Examination of the table shows that at low extraction rates the major part of some of the most significant nutrients is removed, the only gain being an increase in carbohydrate. This would lead us to suspect that the bran, germ and scutellum may be the reservoir of the greater part of the vitamins and minerals in the original wheat (100% column).

Many other studies on the composition of different mill-streams rich in bran or germ tended to confirm this belief, but the definitive experiments

Table 2.2. The effect of extraction rate on the composition of flour experimentally milled from a mixed grist of Manitoban wheats

Constituent	Extraction rate			
	100%	80%	70%	42%
Protein (g/100 g)	13.6	13.2	12.8	11.8
Fat (g/100 g)	2.5	1.4	1.2	0.9
Carbohydrate (g/100 g)	63	69	70	71
Fibre (g/100 g)	2.2	0.13	trace	trace
Vitamin B_1 (μg/g)	3.5	2.0	0.7	0.3
Riboflavin (μg/g)	1.7	0.8	0.7	0.5
Nicotinic acid (μg/g)	55.0	11.0	8.4	7.0
Calcium (mg/100 g)	27.6	15.4	12.8	11.1

were carried out by Hinton who carefully hand-dissected a sufficiently large quantity of wheat into its component fractions to permit analyses of the anatomical components to be carried out. Table 2.3 contains illustrative examples of his data.

The low levels of nutrients (excepting, of course, protein and carbohydrate) in the endosperm, and the special significance of the aleurone layer and scutellum as stores of vitamin B_1, prompt the thought that the miller controls the nutritive value of flour to a marked degree. In the normal European or North American situation this is not of critical significance, since wheat only represents about one-quarter of the total calorie intake, and there are substantial sources of B vitamins and minerals in the remainder. On the other hand, in a siege situation (such as that of the United Kingdom in the second world war) or in circumstances where wheat contributes a major portion of dietary calories, loss of nutrients

Table 2.3. Distribution of B vitamins amongst the principal anatomical portions of the grain expressed as percentages of the total in the grain
Data based on Clegg, K. M. and Hinton, J. J. C. (1958) *J. Sci. Food Agric.* 9, 717–731; and Kent, N. L. (1975) *The Technology of Cereals*, 2nd edn, p. 66, Pergamon Press, Oxford.

Part of grain	Vitamin				
	B_1	riboflavin	niacin	pyridoxine	pantothenic acid
	%				
Pericarp, testa and hyaline	1	5	4	12	9
Aleurone	32	37	82	61	41
Endosperm	3	32	12	6	43
Scutellum	62	14	1	12	4
Embryo	2	12	1	9	3

during mill separations may be undesirable or even dangerous. Thus, during the second world war the extraction rate of flour was controlled by law in the United Kingdom partly to ensure most effective use of limited supplies and partly to ensure that a higher proportion of wheat nutrients reached the human consumer as opposed to finding its way into animal feeding stuffs. In consequence, bread became grey in colour and less attractive to eat, but the medical evidence of the time suggests that health standards did not suffer in consequence. Indeed, in some ways they improved.

As supplies of wheat became more freely available in the post-war years, public demand forced a return to the white bread of pre-war days, and controls on extraction rates were removed in 1953. However, under present regulations all flour in Britain must contain 0.24 mg of thiamin, 1.6 mg of nicotinic acid and 1.65 mg of iron per 100 g. In the case of wholemeal flour these ingredients would be expected to be naturally present but in the case of flours of lower extraction the law requires that they be added. The effect of all this is to ensure that British flour, of whatever extraction, is similar in nutrient content to wholemeal. How far the law really succeeds in this is another matter, since no attempt is made to replace the fibre which at one time was regarded as inert physiologically but which is now known to be of significance in the proper functioning of the large intenstine. Flour is also used to supplement the British diet with calcium, between 235 mg and 390 mg of chalk per 100 g being a legally required addition.

Nutrient addition is made at the flour mill, the nutrients being blended with a diluting agent (usually a low extraction rate flour) in such proportions that the addition of one ounce of this master-mix per 280 lb sack of flour (i.e. 1 part in 4,480 parts by weight) gives the correct degree of supplementation. The addition of the master-mix is made directly to the moving flour stream by specially designed dosing equipment. Enrichment standards not dissimilar to those used in Britain are also applied in the United States and in Canada.

Since flour solids represent 95 per cent or more of bread solids as consumed in Britain, the nutritive value of bread is closely related to that of the flour from which it is baked. In the United States, where milk or milk powder and extra fat are often used in breadbaking, the situation may be more complex but is not essentially different. The process used by the baker also has some influence in terms of nutrients per unit weight as eaten, since modern continuous bread-making processes tend to produce bread of rather higher moisture content than traditional batch processes. Typically the energy value of white bread is around 250–260 kcal/100 g (i.e. 1050–1090 kJ/100 g). For an average slice of 11.5 g, this represents about 29–30 kcal, 120–125 kJ.

Table 2.4. Composition of British bread as prepared by the Chorleywood continuous bread-making process
Data based on Knight et al. (1973) *Br. J. Nutr. 30*, 181

Constituent	per 100 g bread
Moisture (g)	39
Protein (N × 5.7) (g)	8
Fat (g)	1.7
Available carbohydrate (as monosaccharide) (g)	54.3
Ash (g)	1.9
Calcium (mg)	100
Sodium (mg)	540
Potassium (mg)	100
Iron (mg)	1.7
Thiamin (mg)	0.18
Riboflavin (mg)	0.03
Nicotinic acid (available) (mg)	0.82

As eaten in the United Kingdom, bread approximates to the composition shown in Table 2.4.

In this section attention has focussed primarily on wheat in relation to bread making and no mention has been made of its numerous other uses in biscuits, cakes, pie casings and sausage rusks. These uses, of course, are very important but in recent years bread has accounted for a high proportion of the flour usage in Britain and elsewhere, and this has naturally influenced the amount of scientific attention devoted to it.

2.5. The properties of wheat proteins

The standard sack of flour weighs one-eighth of a ton or 280 lb (127 kg). For experimental purposes it is often convenient to work on a similar basis but scaling down so that 1 g is equivalent to 1 lb as used in the factory. If one takes 28 g of a bread-making flour and mixes it with around 16 ml of water, it will form a dough. Now place this dough in the palm of one hand, put it under a gently dripping tap and gently knead it with the forefinger of the other hand. The water will become cloudy with starch washed away from the dough. If this is continued for about ten minutes, the water draining away will gradually clear and the residue left behind is the water-insoluble protein complex known as gluten. The properties of this material provide the key to the characteristics which have given wheat its position at the top of the cereals league.

Of the other cereals rye alone contains gluten but rye gluten characteristics are but a pale reflection of those of wheat. The hybrid of wheat and rye, triticale, also contains gluten but this cereal is more the product of modern plant genetics than of nature. It could not have existed without human intervention and it derives its gluten mainly from its wheat ancestry.

If the experiment previously described is repeated using a biscuit flour, the gluten separated will be quite different in character to that from bread flour. The former has three characteristics which can be readily shown by simple demonstration. First, stretch it by pulling it like a piece of rubber between the two hands. If stretched a short distance it will show signs of springing back when released but not as far back as its original size unless the deformation is small. If stretching is continued until the piece fractures, the broken ends will tend to spring back. Furthermore during stretching, one is conscious that the material is resisting the shearing forces involved in stretching. Thus the material exhibits the capacity for viscous flow of high viscosity combined with a degree of elasticity. Such materials are said to possess viscoelastic properties. The third characteristic is even stranger. If the material is worked by kneading it between the fingers for twenty or thirty seconds and stretched again, its properties change. This time its viscosity is felt to increase by the obviously greater resistance to stretch and it breaks at a much lesser degree of extension than previously—a kind of work-hardening process. If it is kneaded again and then put aside for ten minutes and stretched again, its original properties will have been found to be restored.

Now repeat the experiment with the biscuit flour. It will be found to be much less elastic but will stretch much further before it breaks. The propery of being able to stretch like chewing gum is often referred to as extensibility. Biscuit flours have only a very limited capacity to retain gas. If used for breadmaking, most of the gas produced during fermentation escapes and the resulting loaf has a small, stunted appearance, an unattractive texture and feels heavy and 'doughy' in the mouth. In contrast, bread flours have the right combination of viscous and elastic properties to give good gas retention and this results in bread of bold appearance, good volume and attractive texture and mouth-feel. On the other hand, if bread flour is used for biscuits the dough is too tough and the biscuits shrink in size after cutting-out and before baking. The resulting biscuits are hard rather than pleasantly soft and crisp and, with large automatic biscuit plants, variable shrinkage means variable size at the wrapping machines with papers which don't fit correctly. The resulting chaos on a fast production line will make even the least imaginative mind boggle. Similar considerations apply to other uses of flour such as the baking of cakes, scones and indeed the whole range of bakery products. The flour characteristics must match the in-use product requirements.

These differing flour characteristics are associated with wheats of different origins. Manitoban wheats are regarded as bread wheats, *par excellence*, while English and West European wheats make good biscuits. The descriptive jargon of the baker and miller is confusing on this issue. To start with, the miller tends to describe wheats as hard or soft depending on the characteristics of the endosperm. Some wheats have a hard flinty endosperm which gives clean fractures during milling and fractions easily separated by sieving. Other wheats have a soft mealy endosperm which tends to squash rather than fracture on the mill rolls and which blinds the sieves more readily. Others are intermediate in character.

In general hard wheats are spring-sown and yield flours of bread-baking characteristics. Bakers refer to such flours as strong flours because of their tough elastic gluten. They are equally likely to refer to them as spring flours, since they are derived from spring-sown wheat. On the other hand, winter-sown wheats are often soft in character and yield flours suitable for biscuits, cakes and similar goods. The baker tends to refer to these as winter flours or soft flours. The terms, however, are used loosely and many wheats are intermediate in character between soft and strong. Whatever the origin of the wheat the baker generally refers to bread-making flours as 'springs' and the softer flours as 'winters'. From the farmer's point of view, in climates with mild winters and cool summers, autumn sowing allows the wheat to start growth before winter sets in (winter sowing is a misnomer, the 'winter' wheats are, of course, sown in the autumn). When the temperature of the soil rises to about 8°C in spring, growth starts again and the now established plants grow rapidly and ripen in later summer or early autumn when harvesting conditions are likely to be dry. In countries with severe winters and hot summers such as Canada and parts of the United States, autumn-sown wheat does not survive the winter, but spring sowing is satisfactory with the rapid growth induced by the high summer temperatures.

The baking qualities of different wheats are thus greatly influenced by climate. However, genetic factors also play their part, and wheat breeding is the art and science of matching variety to environment and climate. Farming techniques also play their part. Thus, high protein is generally regarded as a nutritionally desired character and also a technologically helpful one. Yet the full genetic potential of a strain will not be reached if the soil nutrient supply is inadequate. A high-protein strain may give a wheat of only average protein if grown on land insufficiently supplied with nitrogenous fertilizer. Similarly and obviously, maximum yields per hectare will only be achieved by providing soil conditions which supply all the required plant nutrients.

These valuable properties of wheat proteins must be reflected in the chemical structures involved. In 1907 Osborne, that pioneer of protein chemistry, classified flour proteins by solubility as follows

Protein	Percentage (approx.) of total protein	Soluble in
	%	
Albumin	2.5	dilute salt solutions
Globulin	5.0	dilute salt solutions
Protease	2.5	water
Prolamin (gliadin)	40–50	70% alcohol
Glutelin (glutenin)	40–50	dilute acids and alkalis

It was a brave attempt and, if his figures were a bit rough-and-ready, he at least reached the heart of the matter in selecting solubility to provide initial criteria. In modern wheats the albumin and globulin fractions combined would range between 12 and 24 per cent and the gluten fraction (now known to be essentially gliadin plus glutenin together with a small quantity of fat and carbohydrate) ranges around 78–85 per cent of the total protein. But while Osborne's fractions still provide a useful rough classification, it is now known that each is heterogeneous and contains many individual species of protein, some of which cross-link through disulphide bridges during the bread-making process.

The protein chemistry of wheat is still obscure in detail, but as a result of much painstaking work over the past forty years, it is possible to arrive at some generalised idea of the type of structures involved. The high molecular weight fraction of gluten (glutenin) is the fraction responsible for the viscoelastic properties of the dough. Gliadin molecules, which have much lower molecular weights than glutenin, are thought to have a modifying influence on the properties of the latter. An increase of the proportion of higher molecular weight proteins imparts resistance and an increase in strength. Optimum bread-making properties result when the correct balance between these two is obtained.

In a sense, however, conventional concepts of molecular weight are misleading because of the association between the original proteins when they are hydrated. This results in the formation of the three-dimensional protein matrix which forms the giant molecular aggregates in a dough as the baker uses it. Any concepts of molecular structure must account simultaneously for the elasticity, extensibility and response to working (work-hardening and subsequent relaxation with time). The experimental evidence is hard to obtain, since the system exhibits its properties in a hydrated yet insoluble state.

The factual evidence of internal bonding may be summarised as follows.

a) The presence of salts at appropriate concentrations affects dough properties in ways which indicate that ionic bonding plays some role in the giant molecular structure.

b) The action of disulphide-breaking agents, such as *N*-ethylmaleimide, indicates that disulphide bonds are also involved. There is further evidence from mixing studies that disulphide interchange takes place depending on the degree of mechanical working.

c) The action of urea, which breaks hydrogen bonds, is dramatic and evidence on this suggests that huge numbers of such bonds are involved in the cross-linked structure.

d) Hydrophobic bonds (the tendency of non-polar groups to orient towards each other in the same or adjoining molecules) also play their part since acetone, which tends to break such bonds, has a detrimental effect on gluten consistency.

Thus gluten is a system in which several forms of intermolecular and intramolecular bonding are present. On the basis of evidence of this sort together with physical measurements, a number of hypotheses of gluten

Fig. 2.3. Part of a linear glutenin molecule in relaxed form with interchain SS bonds shielded (after Ewart)

structure have been put forward, amplifying and extending the pioneering studies of thirty years ago by cereal chemists such as Blish and Sullivan. Modern views are strongly influenced by Kasarda and Pomeranz but have come to a focus in successive publications by A. D. Ewart during the past twelve years. For detailed information the reader must look to the original iterature.

We now start with a picture of glutenin molecules formed from a linear chain of polypeptides linked together through disulphide bonds, the molecular weight of the individual polypeptides being of the order of 45,000, and the molecular weight of the large molecules so formed being a million or more. It is now thought that these giant molecules are linear in form and can be visualised as polypeptide coils, intramolecularly bonded by disulphide bridges and intermolecularly bonded into chains by the same mechanism as illustrated diagrammatically in Fig. 2.3 and 2.4. The properties of the dough may be enhanced by some degree of mechanical 'entanglement' of these molecules.

During mixing shear forces bring about bond rupture if the mixing is sufficiently severe, and disulphide interchange will take place, perhaps increasing the degree of 'entanglement' (if such a word is appropriate to

Fig. 2.4. Part of a linear glutenin molecule extended under stress with interchain SS bonds exposed (after Ewart).

such bulky structures) at the same time, and certainly providing opportunity for increasing the number of secondary bonds. (At the same time it must be remembered that mixing *per se* is not necessary to produce dough properties. This can be demonstrated quite simply by mixing intimately very finely powered ice with flour at a temperature well below freezing point. When the free-flowing powder mix thus obtained is allowed to rise to room temperature without further mixing, a dough is obtained.)

This model gives a fairly satisfactory explanation of most dough properties. However, it does not tell the whole story since the behaviour of real-life dough is influenced also by the lipids present, by the fermentation process itself and by the fact that starch granules are embedded in the gluten network.

The student coming fresh to this topic would do well to realise that the account given here is only an outline of a complex physicochemical phenomenon. Many years ago the late Professor James P. Todd remarked to this writer that 'the scientific phenomena underlying the production of a loaf of bread are much more complex than those underlying the design, production and operation of a nuclear reactor.' The passing of the years has done nothing but enhance the validity of this casual comment.

But the student may well ask 'why bother to attempt to understand the detail of such a complex system when we know perfectly well how to make good bread?' There are two answers. At one level the simple curiosity which is the ultimate mainspring of scientific activity is sufficient. At a more practical level, the fuller our knowledge the better we are placed to use wheat, that most useful of all the cereals, to the best advantage. There is almost certainly still more to learn about it than the sum total of all our modern knowledge.

2.6. Maize

Maize or Indian corn (*Zea mays*) is essentially a tropical or subtropical crop but it has spread gradually northwards through Europe and varieties suitable even for the mild climate of the South of England are now available and are being commercially grown. It is widely used for cattle and poultry feed but significant proportions go to direct human consumption as breakfast cereals, cornflour and corn syrup (known in the U.K. as confectioner's glucose) and indirectly as potable spirits. In Mexico and elsewhere it is also used to produce unleavened bread in the form of tortillas. In its listing in the annual tonnage league it is second only to wheat. More than half of the total world crop is grown in the United States, and about three-quarters of this is fed to livestock.

The origin of maize is unknown and it is now only found under cultivation. The primary centre of origin of the plant is most probably central America where the similar wild plant teosinte (*Euchlaena mexica*) still flourishes. The two genera can be hybridised readily. In his log of November 6th, 1492, Columbus reports having seen large quantities of a grain which the Indians called 'maiz', and later explorers found maize to be in native cultivation everywhere from what is now southern Canada to southern Chile. There is archeological evidence of the cultivated plant having been grown in New Mexico about 5,600 years ago (carbon dating of the cobs). Maize was introduced to Europe in 1494 after the return of Columbus from his second voyage and within a few years had spread to southern France, Italy and North Arica.

Maize is commonly classified into six types: dent, flint, flour, sweet, pop and waxy. These subspecies are fully interfertile. The sweet corn has a high sugar content and much is harvested when tender and immature for canning and freezing. The dent type is the major commercial variety grown in the United States and is characterised by indentations in the side of the mature grain and by its hard vitreous endosperm. Flint corn does not show this indentation and has some soft starch at the centre of the grain while flour corn was a kernel consisting largely of soft starch. Popcorn has a high percentage of hard endosperm which makes it burst open on heating. The endosperm starch in waxy corn consists almost entirely of amylopectin which gives it a wax-like appearance on cutting. It is valued for special industrial uses such as the manufacture of adhesives.

Hybrids obtained by crossing two or more inbred lines of corn are now very widely grown in many parts of the world for their high yields and adaptation to environmental conditions, some lines of hybrids also being better able to withstand diseases, drought, insects and other adverse environmental factors. As a result of improved varieties and agricultural practices, average maize yields in the United States increased from 2.2 to 5.3 tonnes per hectare between 1945 and 1970.

The anatomical proportions of the grain vary a little from type to type but typically would be as follows

Pericarp	7–8%
Endosperm	76–82%
Embryo (including scutellum)	10–12%

Similarly the principal chemical constitutents of the kernel vary with type, variety and growing conditions as shown in Table 2.5.

2.6.1. Maize proteins

Zein, the principal protein in maize, is deficient in both tryptophan and lysine but maize glutelin, the other important maize protein, is of rather

Table 2.5. The chemical composition of the maize kernel.
The figures for vitamins and minerals are to be regarded as typical. There is of course, some variation

Constituent	per 100 g maize kernel
Moisture (g)	10.8–13.5
Protein (g)	9.3–10.0
Fat (g)	4.0– 4.4
Fibre (g)	1.7– 2.3
Carbohydrate (g)	68.1–72.0
Ash (g)	1.3– 1.5
Calcium (mg)	10
Phosphorus (mg)	26
Iron (mg)	2.3
Thiamin (mg)	0.45
Riboflavin (mg)	0.10
Niacin (mg)	2.2
Pantothenic acid (mg)	0.6
Pyridoxin (mg)	0.6
Vitamin E (mg)	2.5
Vitamin A (μg)	140–900 (yellow varieties only)

better quality. High lysine mutants known as opaque-2 and floury-2 were identified in 1964 and from them, high-lysine varieties of corn have been commercially available since 1970.

2.6.2. Vitamins in maize

Pellagra is a disease prevalent in areas of poor diet where maize is the staple food and accounts for most of the calorie intake. Pellagra is due to a deficiency of dietary niacin (see page 31). While maize appears to contain adequate amounts of this vitamin, much of it is present in a bound form, known as niacytin, which is biologically unavailable. It is interesting to note that alkali releases the bound niacin. Mexican tortillas are made traditionally by mixing maize meal with lime water. In Columbia potash is used to loosen the bran when the grain is pounded in the domestic preparation of maize meal. Both of these processes have the effect of releasing the bound niacin and the users rarely suffer from pellagra.

2.6.3 Maize oil

The embryo of the maize kernel contains about 30 per cent oil. During wet milling of maize the embryo is separated by flotation and the oil is extracted by the use of high-pressure oil expellers or by solvents. The oil has excellent cooking qualities, and its high levels of linoleic and linolenic acids are valued in dietary after-care of patients suffering from circulatory degeneration.

It is a particularly good oil for cooking potato chips or French-fried potatoes. It has a quite bland flavour and does not mask the natural flavour of foods cooked in it.

2.6.4. Uses of maize

The United States produces about 58 per cent of the world's crop and uses about three-quarters of this as animal fodder, and in world terms maize must therefore be regarded primarily as a feed crop. However, an appreciable and important fraction is used for human food either directly (as already described in Mexico or Columbia) in the form of meal or indirectly in the form of products such as corn starch, corn syrups, corn flakes and bakery products. A major use is in the production of alcoholic beverages both in the United States and elsewhere. Proprietary brands of Scotch whisky are blended from grain and malt-based spirits, the grain used being maize whose starch is converted into the sugars required for yeast fermentation by the action of diastatic enzymes present in the barley-malt added to the extract (technically the 'mash') prior to fermentation. Butyl alcohol and acetone are also produced commercially by bacterial fermentation of corn.

2.7. Rice (*Oryza sativa*)

2.7.1. Introduction

Rice is a swamp cereal, requiring moisture and heat. It comes third to wheat and maize in terms of world cereal production but, nevertheless, is the principal food crop of about half of the population of the world. The principal growing areas are in India, Bangladesh, China, Japan and South-East Asia, but smaller quantities are grown in Italy, Egypt, Spain, the Philippines, Brazil, Portugal, Taiwan, Chile and the United States. In China its culture goes back at least 5,000 years, and the cultivated rice plant is believed to have originated in the area between India and Indochina. Wild forms of rice are still to be found in this area. Most of the rice crop is grown on artificially or naturally flooded land, which must obviously be level or carefully terraced, but some is grown at higher altitudes in situations where the rainfall is sufficient. The growing season is from four to six months and a mean temperature of at least 21°C is required throughout the growing period. Most of the world's rice is grown where the annual rainfall is 100 cm or more.

More than 8,000 botanically different rice varieties are known, but by and large they fall into two main types of groups. The *indica* varieties are mostly long-grained, while *japonica* strains are shorted-grained. In addition *Zigania aquatica*, a form of wild rice, is harvested in small quantities in certain areas where it is prized for its non-glutinous aromatic character. In general, the short grained varieties are better adapted to cooler climates and are reputed to respond better to heavy dressings of fertilizers. Harvesting can be highly mechanised, but in Asiatic countries much rice is still harvested using hand sickles followed by hand threshing. After threshing the grain is known as paddy rice (sometimes referred to as rough rice). It consists of about 20 per cent hulls and 80 per cent grains. The hulls consist of the lemma and palea, two leaf-like structures enclosing the original flower. The first stage of milling removes them to release the caryopsis, or kernel.

This is known as 'brown' rice or 'husked' rice. Thereafter the bran layers are removed by abrasion and finally the now naked grain is polished in a machine known as a brush machine to produce the product of commerce.

2.7.2. Chemical composition

A typical analysis of rice at the three principal production stages would be as shown in Table 2.6.

The figures will vary with location and variety but the general pattern remains the same. The hulls represent about 20 per cent of the grain as harvested. Their utilisation is handicapped by their low nutritive value, their abrasive character, their low bulk density, their resistance to degradation and high ash content, which is around 20 per cent and consists mainly of silica. Their disposal is often a serious problem to the rice miller. They may be burned to generate steam but the high ash content merely produces another disposal problem. They have been used as chicken litter, as a raw material for the manufacture of furfural and hardboards, as a filter aid, in rubber manufacture and even as metal and gem-stone abrasives. Despite the high level of investigation into finding

Table 2.6. Compositional changes arising from rice milling.

Constituent	Paddy rice	Brown rice	Polished rice
	%		
Moisture	11.3	12.3	11.8
Protein	7.5	8.6	8.1
Fat	1.6	1.8	0.3
Fibre	8.7	1.0	0.3
Carbohydrate	65.5	75.1	79.1
Ash	5.4	1.2	0.4

Table 2.7. Composition of rice bran

Constituent	%
Moisture	11
Protein	12
Fat	16
Crude fibre	12
Ash	10
Nitrogen-free extract	39

uses for rice hulls, their disposal remains something of an embarrassment in areas of intensive rice production.

In contrast, rice bran is a nutritionally valuable commodity of variable composition but rich in protein, fat, minerals and vitamins. Table 2.7 gives a representative analyses.

These figures are taken as typical of an average sample based on an examination of a number of analyses reported at different times from different sources. The student will appreciate that the composition of bran will depend to a marked extent on the severity of milling (i.e. whether some endosperm is removed with the bran), on the effectiveness of the preceeding hulling and separating processes (inclusion of hull fragments) as well as on variety and growing conditions.

2.7.3. Vitamins in rice

Beri-beri, a disease endemic in rice-eating communities, was first shown to be of dietary origin in 1884 when Takaki eradicated it from the Japanese navy by an alteration in rations. The discovery of dietary polyneuritis by Eijkman in Java in 1896 and the proof by Frazer and Stanton in the then Malay States of the cause of beri-beri by a diet of polished rice, and its relief by substitution of unpolished rice, initiated a search for the chemical substance in rice polishings which prevented the disease. The substance was, of course, thiamin and it was eventually isolated in 1926 by Jansen and Donath. Thus, without prior treatment, rice polishing removes most of the vitamin B_1 from the grain, as well as the other B-group vitamins and much of the minerals.

The parboiling of rice has long been practiced in the Indian subcontinent as an aid to milling. The process has many variations but in principle it consists of soaking the paddy rice in hot water followed by steaming and then drying to a suitable moisture content before milling. As well as loosening the hulls, the nutritive value of the milled rice is enhanced by this treatment, which removes some of the water-soluble minerals and

Table 2.8. Effect of parboiling of rice on nutritional value

Constituent	Brown	Milled	Milled parboiled
	per 100 g		
Water (g)	12.0	12.0	10.3
Protein (g)	7.5	6.7	7.4
Fat (g)	1.9	0.4	0.3
Ash (g)	1.2	0.5	0.7
Total carbohydrate (g)	77.4	80.4	81.3
Fibre (g)	0.9	0.3	0.2
Calcium (mg)	32.0	24.0	60.0
Phosphorus (mg)	111.0	94.0	200.0
Iron (mg)	1.6	0.8	(2.9)[a]
Sodium (mg)	9.0	5.0	9.0
Potassium (mg)	214.0	92.0	150.0
Thiamin (mg)	0.34	0.07	(0.44)[a]
Riboflavin (mg)	0.05	0.03	
Niacin (mg)	4.7	1.6	(3.5)[a]

[a] Enriched.

vitamins from the hulls and bran-coats which migrate to the endosperm. Parboiling toughens the grain and reduces breakage on milling. It is also said to reduce susceptibility to insect attack.

The nutritional advantages are illustrated in Table 2.8.

The early Indian processes, developed by some unsung village food technologists in a bygone century, consisted of filling large metal or earthenware pots with paddy rice and water and setting the pots on a thick bed of rice bran which was then set on fire. The fire burned for many hours and afterwards the parboiled rice was dried in the sun.

As now adopted in large rice mills, the process is a carefully controlled sequence of hot-water treatment, (sometimes after application of a vacuum to speed up water penetration) followed by steaming and drying. The drying ratios are critical if the best quality is to be obtained, and the overall operations are conducted using modern large-scale equipment whose design is based on sound chemical engineering principles.

The advantages of parboiling are such that the process has spread from India to many other parts of the world. However, its acceptance is far from universal for the following reasons. Parboiled rice develops more rancidity during storage than does raw rice, this being attributed to destruction of natural antioxidants during parboiling. Parboiled rice takes longer to cook and often has a distinctive flavour and colour disliked by some raw rice eaters. Parboiled rice is more difficult and expensive to polish than raw rice. Parboiling requires more capital in machinery than do the more traditional forms of milling.

An alternative approach to the nutritional enrichment of polished rice by parboiling, is the addition of amino acids, minerals and vitamins. This is by no means a simple problem compared, for example, with the enrichment of white flours made from wheat. In most countries rice is washed (often several times) before cooking. Sometimes it is boiled in excess water and the surplus discarded, and with either procedure any superficial addition of nutrients will be lost.

The American approach is to sprinkle the enrichment solution on to the rice in a slowly rotating cylinder in the presence of a simultaneous stream of air to remove moisture. After the completion of this process the nutrients are sealed in with the addition of an alcoholic solution of stearic acid, zein (a protein extracted from maize) and abeitic acid. The finished rice is finally dusted with ferric pyrophosphate and talc.

The Japanese approach to the same problem is based on the production of water-insoluble derivatives of the active ingredient using 1 per cent acetic acid (sometimes containing an organic solvent) as the mode of application of the fortifying nutrients to the kernel. Thiamin, together with lysine and threonine, can be applied in this way, and once applied are not readily leached out by subsequent cooking. The process is used to produce a premix concentrate of the added nutrients, which is then diluted by normal rice at the rate of one in two hundred. There are a number of variations in this process developed originally by Mitsuda and it is estimated that about one-third of the Japanese population benefits by the consumption of enriched rice.

2.8. **Barley** (*Hordeum species*)

2.8.1. Introduction

Barley is perhaps the most tolerant of all cereals to a wide range of climatic and environmental conditions. It is grown from the Arctic Circle to the tropics and is even adaptable to high altitudes, useful crops being obtained at heights of up to 4,500 m in the Himalaya mountains. It is one of the most dependable crops where frost, drought or alkaline soil conditions are likely to be encountered. It grows particularly well in climates with a long cool ripening season.

Wild forms of barley are found in various parts of the world and the source of our present cultivated varieties is not known with any degree of certainty. Barley bread was used by the Egyptians (some survives in museums to this day) and the abundance of wild varieties in the Middle East lends support to the idea that cultivation may have begun there.

During the middle ages barley/rye or barley/wheat mixtures were used by the peasant and poorer classes in England as a bread cereal, bread made from wheat alone being the prerogative of the wealthy. Bread made from mixed cereal grists is generally rather course, heavy and unappetising. The use of barley in breadmaking has long since disappeared in Europe, although during the second world war small quantities of barley flour (up to a maximum of 10 per cent) were added to extend the limited supplies of wheat available.

2.8.2. Uses

Only small quantities of barley are used directly for human food. The main part of the world's crop goes for the production of malt or for animal feed. Barley malt is the key raw material in the brewing and distilling industries, and is prepared by soaking the grain to allow germination, followed by kilning (shortly after the first rootlet or acrospire appears) to arrest the process. The starches in the endosperm of the grain are thus largely converted to fermentable sugars and the amylase activity of the grain is greatly increased, which in turn plays its part in some subsequent fermentation processes.

Small quantities are used directly in human food in the form of barley flour, which can be quite a useful thickener for sauces and soups and is also used in some breakfast cereals. Pearl barley is used in soups and in a few recipes for comminuted meat products, such as Scotch haggis, black puddings and similar products. Barley water and barley groats are sometimes used in infant feed to prevent a thick difficult-to-digest curd from forming in the infant's stomach, especially when cow's milk is fed. Barley water is claimed to have a cooling effect on fevers and a soothing effect on certain digestive disorders, but its usefulness for these purposes is not related in any known way to its composition.

2.8.3. Chemical composition

When barley is to be used for animal feeding, a high protein content is desired. For malting, a high carbohydrate content is sought. As a result protein contents ranging from 7.5 to 15 per cent on a dry-weight basis have been reported in the whole grain, with starch figures ranging from 50 to 60 per cent. Since the protein content rises with applications of nitrogeneous fertilizers, application of these to barleys for malting must be restricted.

In 1905 Osborne separated barley proteins into the following fractions based on their solubility:

Leucosin (an albumin)	3%	(water soluble)
Edestin (a globulin)	18%	(dilute salt solutions)
Hordein (a prolamin)	38%	(alcohol soluble)
Undesignated (a glutelin)	41%	(alkali soluble).

Although it is now known that each of these contains several component proteins, the classification remains useful for practical purposes to this day. The proportions of each fraction will vary from variety to variety and also with cultivation conditions. Recent work by Brandt has shown that after fertilisation these different fractions accumulate in the developing grain at different rates, the albumin and free amino acid fraction developing very rapidly while the hordein and glutelin fractions appear gradually at first and only reach their peak about twenty-eight days after fertilisation, by which time the grain is approaching ripeness.

Like other cereals, barley protein tends to be low in lysine. Hiproly (i.e. hi-pro-ly) barley is a mutant from Ethiopia which contains a high-lysine gene. One of the strains to date is a developed variety from Denmark known as Riso 1508 which has up to 50 per cent more lysine than ordinary barley. Recent work (1976) has shown that the high lysine content of the mutant endosperm is due to a reduction in the lysine-poor hordeins, a reduction in the lysine-poor components of the glutelin fraction, and a compensating increase in the amount of lysine-rich glutelins as well as in free lysine.

As with the other cereals, whole barley grains are a useful source of B-group vitamins, the general levels of the main members of the group being similar to those in wheat. However important these may be in animal feeding, barley for direct human use is usually in the form of pearl barley or barley flour and the removal of the seed coats and germ in making these products results in much lower vitamin levels than those present in the whole grain. However, barley does not play an important part in the human diet except in the form of beer where thankfully some of the B-group vitamins are retained.

The analysis in Table 2.9 is typical of the dehusked barley grain.

As with all such tables the figures should be regarded as merely representative. In practice a range between samples is to be expected.

Table 2.9. Composition of dehusked barley kernels

Constituent	per 100 g
Moisture	14.0 g
Protein	10.6 g
Fat	2.1 g
Starch etc.	66.0 g
Fibre	4.5 g
Ash	2.5 g
Thiamin (B_1)	6.5 μg
Riboflavin	1.2 μg
Nicotinic acid	90.0 μg
Pantothenic acid	4.4 μg
Pyridoxine	3.5 μg

2.9. **Oats** (*Avena sativa* and related species)

In Samuel Johnson's great eighteenth century Dictionary of the English Language is the following entry

Oats: A grain, which in England is generally given to horses, but in Scotland supports the people.

Boswell in reading the entry looked at him and said 'Where, sir, will you find better horses than in England or better men than in Scotland?'

It is certainly true that oatmeal was a staple food in Scotland for centuries, the cool moist climate favouring it rather than wheat which fared better in the warmer south. In modern times it has come to be used mainly for feeding livestock and probably not more than about 5 per cent of the world's crop goes for human food. This is a pity, since when properly prepared and cooked, oats form the basis of such excellent and nourishing dishes as oatcakes and porridge, and are used also in local recipes such as black puddings and haggis. Oatcakes and butter, porridge and milk, and black pudding or haggis with potato and turnip or greens make platters which are nutritionally well balanced and tasty to boot.

The husk of the wheat grain separates readily as chaff during threshing, but like that of rice the oat husk is firmly adhering and this affects the milling process. On the other hand, compared with wheat, the bran of the oat kernel is thin and pale in colour, and it is not, therefore, necessary to separate it when making oatmeal or rolled oats. At the same time the oat kernel contains 2–5 times as much fat as that of wheat, and its pericarp is quite rich in lipase. This affects the processing methods used, which aim to minimise the risk of rancidity on storage. Oat proteins do not form an adherent gluten and oatmeal or flour is therefore unsuitable for bread-making.

Oats are thus commonly used in the form of groats (fragments of kernel) or oatmeal. The bulk grain from the farm is first cleaned by processes similar to those used in flour milling. The cleaned grain is thereafter treated by injection of live steam to raise the temperature to about 95°C to inactivate the lipase, which may cause rancidity in oatmeal or the development of soapy flavours in products made from it. Thereafter it is kilned by passing it down vertical towers through which the combustion gases from a coke or anthracite furnace are passed. Kiln drying reduces the moisture content to 3–4 per cent which facilitates the subsequent dehusking process by making the husks brittle. It also improves the flavour, the grain developing a nutty taste in the process. The grains then pass between two large stone millwheels, separated by a distance slightly less than the length of the grain, one stone being fixed while the other rotates. The husks so cracked and loosened are removed

Table 2.10. The chemical composition of oatmeal
The figures quoted are for a characteristic sample. As with all cereals, variation from batch to batch is to be expected.

Constituent	per 100 g
Moisture (g)	8
Protein (g)	14
Fat (g)	7
Fibre (g)	1
Carbohydrate (g)	68
Ash (g)	1.7
Calcium (mg)	55
Phosphorus (mg)	380
Iron (mg)	4
Thiamin (mg)	0.6
Riboflavin (mg)	0.1
Niacin (mg)	1.0
Pantothenic acid (mg)	1.0
Pyridoxine (mg)	0.12

by aspiration and further stages of stone-grinding and sieving follow from which oatmeal of varying particle size is produced as well as oatflour.

Rolled oats are widely used for making porridge because of their ease of cooking. The groats are steamed, passed between heavy rollers and then dried, thus producing flakes which cook quickly.

The nutritive value of oatmeal is not dissimilar to that of the other cereals except for the markedly higher fat content. It is a satisfactory source of vitamins of the B group (similar to wheat but richer in B_1 and poorer in nicotinic acid) and its contribution of dietary minerals is similar to that of wheat (see Table 2.10). Oat protein is somewhat richer in lysine and lower in glutamic acid than wheat, but the differences are not such as to be of practical importance in a normal mixed diet.

2.10. Rye

Rye (*Secale cereale*) was at one time, and to a degree still is, the principal cereal of Eastern and Northern Europe. It is a favourite bread grain of Scandinavian countries. In the Middle Ages its use was widespread throughout Europe but it has long since been displaced in the more southerly countries whose softer climate permits the growth of wheat. In the Middle Ages a mixture of rye and wheat was used to make maslin, a kind of bread favoured by the poorer people in England.

Rye grains are long and narrow compared with the plump grains of wheat. Rye tolerates poor, acid or sandy soils better than wheat and its deeper-rooting habit gives it better drought resistance. It is intolerant to heat but winter-sown rye easily survives temperatures too low for wheat production, winter types resisting exposures of −30°C. Flour made from rye is dark in colour and although rye proteins have some gluten-like characteristics, these are inferior to those of wheat. Rye bread is therefore both dark in colour and heavy and dense in texture compared to wheat bread. It is often baked by a sour-dough process in which a natural lactic acid fermentation is encouraged using a starter culture based on sour milk. It is also used to bake the traditional Swedish crispbreads and much of the rye consumed for human food in southern Europe is prepared in this way.

About 220 million hectares of land are used in producing the world's wheat crop while rye occupies only about 20 million hectares. It is thus a relatively minor crop in terms of its significance in human nutrition. Despite its poor baking characteristics, rye does have the important nutritional advantage of a (relatively) high-lysine protein.

A good quality malt can be produced from rye which is the raw material of rye whisky. It finds some uses in breakfast cereals, in biscuits in the form of rye flour, and as a filler for sauces, soups and custard powders. It is also used as an animal feed. A typical analysis of rye grains is shown in Table 2.11.

Ergot (*Claviceps purpurea*) is a fungus disease of rye of particular importance because of the toxic nature of the sclerotia produced. These contain alkaloids poisonous to livestock and humans, and rye is considered dangerous for feed or food when it contains more than 0.3 per cent of ergot sclerotia by weight. The use of ergot-free seed and proper systems of crop rotation form the most effective methods of control, resistant varieties being unknown.

The sclerotia of *Claviceps* range from 10 to 20 mm in length and when dry are dull grey or purple-grey in colour. Ergotism in man is of two types, convulsive and gangrenous. Convulsive ergotism is associated with

Table 2.11. Composition of rye grains compiled from a range of published sources

Constituent	per 100 g
Moisture (g)	12–16
Protein (g)	8–12
Fat (g)	1.6–1.9
Ash (g)	1.7–2.0
Fibre (g)	1.7–1.9
Calcium (mg)	32–50
Iron (mg)	2.7–3.5
Vitamin B_1 (mg)	0.3–0.5
Nicotinic acid (mg)	1.2
Riboflavin (mg)	0.10–0.29

the consumption of ergot in conjunction with a deficiency of vitamin A. It is characterised by fatigue, giddiness, gastro-intestinal disturbance, numbness in the limbs and, in severe cases, convulsive attacks. Gangrenous ergotism is caused by high levels of ergot consumption, the initial symptoms being swelling of the limbs, extreme discomfort of the skin (described by patients as a feeling that insects are crawling over them) followed by violent burning pains and dislocation of circulation to the point of limbs becoming useless and mummified. Gangerous ergotism was known as 'St. Anthony's Fire' or 'Holy Fire' in the Middle Ages when it was much more common than at present.

The active alkaloids in ergot are ergotinine and ergotoxine and they find uses in medicine and especially in midwifery for inducing contractions of the womb.

2.11. Triticale

Triticale is a hybrid of wheat and rye and derives its name from the respective generic names of the parent plants. Cross a horse with a donkey and a sterile mule results. Intergeneric hybridisation of cereals is known to occur by chance when fields of different cereals adjoin but the resulting hybrids are infertile.

The earliest attempt to produce a triticale were reported by Stephen Wilson as far back as 1874 in the *Transactions of the Botanical Society of Edinburgh* and from the description given, it seems likely that he did produce the first triticale. There was little progress made during sporadic experimentation over the next fifty years, until the introduction of colchicine treatment for inducing the chromosome doubling of young hybrid plants in 1937 opened the possibility of a fertile result. Progress remained slow until the late nineteen sixties when cultivars became available for trial by commercial farmers.

From the agronomic viewpoint triticale will grow and yield well on land more suited to the cultivation of rye than wheat. It is more resistant to cold and drought and appears to compete well with its rivals in high uplands and on rather poor sandy soils. Yet on comparable land and in suitable environments it is capable of slightly out-yielding wheat. The possibilities of genetic improvement are still substantial and its is expected to make a useful contribution to world cereal resources in the future.

Nutritionally it has advantages over wheat in that it has a higher average protein content of about 1 per cent and, because of the high-lysine gene inherited from its rye parent, the protein is of improved biological value.

From the viewpoint of the cereal technologist, triticale is a better bread-making cereal than rye but it in no way compares with the best wheats. However, this may not be of great importance since its value may lie mainly in areas unsuited to wheat and where leavened bread is not an important part of local diets. However, it has been shown experimentally to produce a range of products (pasta, chuppattis, biscuits, cakes and even breakfast cereals) of quite satisfactory general quality. In addition, the rye parent carries a pleasant flavour (usually described as 'nutty') into baked goods made from it, conferring an added advantage.

Preliminary experiments also indicate that it malts well and that satisfactory beer can be brewed from it. It has also found application as an ingredient in rations for poultry, pigs and ruminants.

2.12. The sorghums and millets

Grain sorghum (*Sorghum vulgare*) is a coarse cereal cultured extensively in Africa, China, India and to a lesser extent in Iran, Pakistan, Korea and the United States. The history of its cultivation is unknown, but it may have been one of the earliest domesticated plants. It is essentially a tropical plant which may be grown in areas too dry and too hot for maize, and will crop better than maize in marginal conditions of uncertain rainfall where maize can be grown but only with difficulty. It is therefore well suited to the semi-arid tropics. To many consumers, however, maize is the more attractive crop to eat, and tropical plant breeders are now giving attention to selecting sorghums for improved eating quality characteristics rather than for yield alone.

The plant is a coarse annual with stems ranging from 60 to 300 cm in height and occasionally even more. The grain head forms a panicle from 7 to 50 cm in length and 5 to 20 cm wide, the seeds being carried on branch-like growths separated from the main stem.

In the United States most of the crop goes for animal feed but about 10 per cent is used industrially in the manufacture of starch, with a protein feed and edible oil as byproducts.

In many African countries such as the Sudan, Ethiopia, Uganda, West Africa and South Africa, the grain is used for human consumption in a variety of forms. Porridge and gruels may be made from sorghum flours, types of pasta products and sorghum pancakes are also liked as well as flat cakes and an unleavened bread. Sorghum is often mixed with other products such as cassava and wheat. The grain can be readily malted, and in Uganda malted sorghum is used to make a form of sweet porridge. In South Africa the so-called Bantu beer is made from sorghum malt by a mixed fermentation. It is a cloudy liquid containing some alcohol and has a pleasant refreshing taste. It represents an important source of nutrients (especially B-group vitamins) for the peoples who use it.

The sorghums are similar in general composition to the other cereals.

The word millet is used to describe a range of species of Gramineae of rather confusing botanical origin. Generic names include *Panicum, Setaria, Elusine, Pennisetum, Paspalum, Eragrostis* and *Aeroseras*. The confusion is worse confounded by the range of local names used to describe millets such as 'hog', 'proso', 'kodo', 'haraka', 'bajoa' and 'teff'.

They are warm-weather annual grasses growing up to 120 cm in height and the grain is small in size containing 10–12 per cent protein and 3–5 per cent fat. Millets are grown in the United States mainly as a feed grain but are widely used for human consumption in other parts of the world. In the U.S.S.R. it is eaten as a thick porridge called kasha, or in the form of a flat bread.

Not dissimilar uses are found for the plant in other parts of the world. In India the unripe ears of finger millet (*Eleusine coracana*) are eaten as a vegetable and the ripe grain is sometimes used for brewing beer. Some millets in India are boiled with milk and sugar. In other parts of the world the meal is cooked as a gruel or porridge and eaten with a sauce or stew. Millet is used also in West Africa for making couscous, fritters and (in combination with wheat, usually 30 per cent millet + 70 per cent wheat) as a bread cereal.

In a word, the millets provide a wide range of qualities and properties and with the ingenuity of the local user, an equally wide range of quite attractive and nutritious staple foods can be prepared from them. While not approaching the significance of wheat, maize and rice in world terms, millets are important cereals in the semi-arid tropics and contribute to food supply in many third world countries.

Further reference

Technology of Cereals (1975) Kent, N. L., 2nd edn, Pergamon Press, Oxford.

Modern Cereal Chemistry (1967) Kent-Jones, D. W. and Amos, A. J., 6th edn, Food Trade Press.

Wheat Chemistry and Technology (1971) Pomeranz, Y. (ed.), American Association of Cereal Chemists, St. Paul, Minnesota.

Rice Chemistry and Technology (1972) Houston, D. F. (ed.), American Association of Cereal Chemists, St. Paul, Minnesota.

Rice—Post Harvest Technology (1976) Araullo, E. V., de Padua, D. B. and Graham, M. (eds), International Development Research Centre, Ottawa.

Rice (1970) Grist, D. H., 4th edn, Longman, London.

Barley and Malt—Biology, Biochemistry, Technology (1962) Cook, A. H. (ed.) Academic Press, London.

Sorghum and Millets (1977) Dendy, D. A. V. (ed.), Tropical Products Institute, London.

Cereal Crops (1963) Leonard, W. H. and Martin, J. H., Macmillan, New York.

3

Meat

3.1. Introduction

A biologist from space observing the behaviour of mankind would find meat-eating a curious phenomenon. He would note that, *Homo sapiens sapiens* apart, the sixty-odd genera of primates were essentially vegetarian. True, most primates will eat the occasional insect with relish, some raid birds' nests for eggs or fledglings, others living near water will trap crabs and the like and baboons and chimpanzees will even hunt small animals for food, but these are dietary luxuries which do not make an important contribution to total protein intake. Thus, an adult baboon weighing around 40 kg will average around 1 g of animal protein per day, while a typical specimen of *Homo sapiens sapiens* weighing 65 kg will consume up to 40 g and more in a free-choice situation.

Were he to pursue the matter further by tracing back the ancestral line of this odd genus, our biologist from space would be yet more surprised to find that its evolutionary ancestors were vegetarians, and that carnivorous man first appeared on the fossil record a mere one and a half or two million years ago. In evolutionary terms, this is recent stuff—one or two megayears in the 70 megayear span of primate evolution. As a carnivore, man is still a baby.

This is curious enough, and as far as I know has no parallel in the evolutionary story. Yet, its consequences are yet more curious. A raw vegetarian diet requires dedicated chewing with teeth and jaws to match. Try it for yourself by eating a couple of blades of grass and you will find that you have a great deal of chewing to do before you are ready to swallow. On such a diet most of the daylight hours must be devoted to foraging and chewing, and the life characteristics which ensue are reflected in the daily routine of, for example, the modern gorilla. Raw meat is a much more concentrated source of energy and protein than raw vegetables, fruit and roots. The evolutionary transition to meat-eating was therefore associated with gradual physiological adaptations in the forms of a smaller jaw (which, some have held, allowed a marked development of the frontal lobes of the brain), smaller teeth, and less demanding attachments of the jaw musculature. In terms of life mode, hunting skills must have been deployed and there is evidence, in the kitchen rubbish left by these remote kinsmen, of the development of simple weapons which they used as clubs to kill their prey. We need not concern ourselves here as to how or why the first transition from a vegetarian to an omnivorous diet took place, but the fact that it did poses problems to modern societies.

Man is well-adapted to a wide-ranging dietary pattern. He can live in excellent health on a diet completely free from animal foods of any sort. Such a strict vegetarian diet is normally devoid of vitamin B_{12} (unless it contains dead insect fragments, mouse droppings or certain fungal contaminants!), and the vitamin must be added artificially from an external source. However, few people enjoy such a diet on grounds of its culinary attractiveness. Most strict vegetarians (vegans) adopt it for ethical or religious reasons. Amongst poor communities, diet may be largely vegetarian through financial circumstances rather than choice, vegetable foods being usually much cheaper than meats. However, when increasing income extends range of choice, dietary intakes of meat and fish increase. Of course, there are individual exceptions, but the trend in populations is clear and overrides race and culture.

The implications are conveniently illustrated by the well-known comparison

1 hectare of soya beans will sustain a man for 5560 days
1 hectare of wheat will sustain a man for 2190 days
1 hectare of corn will sustain a man for 885 days
1 hectare of beef will sustain a man for 192 days

In a world of rapidly increasing population and with a fixed land area the message of the comparison is obvious enough. Yet it is inadequate, because it does not compare like with like. In many parts of the world, cattle can graze where conventional cropping is either impossible or uneconomic. Cattle can convert plants inedible to humans into meat and milk, and on these grounds alone their future seems assured. But if population pressures continue to rise, the use of arable land for cattle may tend to decline and over-wintering feeds (such as fish meal and protein concentrates) will in part find their way to direct human consumption. Gradually diminishing supply may meet increasing demand from the growing wealth of the less developed countries.

For all these reasons the very human desire to eat meat may be gradually modified by a combination of rising demand and falling supply. The development of meat analogues is therefore doubly welcome. As technological skills in their preparation grow, their quality will improve, but already they provide useful service as extenders of comminuted meat products. Furthermore, as already suggested, meat-eating seems inbuilt in man's evolutionary inheritance. But this is so recent a development that a return to an essentially vegetarian diet does not present serious physiological or nutritional problems. On the other hand, any compulsion in this direction will produce social stresses which may be partly relieved through the use of such analogues. This then is the background against which the student approaches the study of meat and meat products.

3.2. The meat animal

Meat is one of the most variable in composition of all the raw materials of the food industry. The main variables are as follows.

a) The type of meat: beef, mutton, pork, horse, venison, goat, fowl etc.

b) The breed of animal: for example cattle bred for meat, such as Herefords and Aberdeen Angus, differ from milk breeds such as Ayrshires or Jerseys.

c) The age of the animal; in general, older animals make tougher eating than younger ones.

d) The amount of exercise the animal has taken: wild animals foraging for their own food are usually darker of flesh, stronger of flavour and tougher of texture than domestic animals.

e) The nutritional plane of the animal: that well-fed animals are usually fatter than ill-fed ones, goes without saying. It is less obvious that the distribution of fat and muscle varies with the nutritional history. Identical twin cattle for example, slaughtered at the same age would differ greatly in composition if one had been underfed during the first three months of its life and thereafter well-fed while the other had been well-fed in infancy and underfed for the last three months of its life.

In addition to variations from animal to animal, there are also variations in chemical composition and eating quality from one group of muscles to another, and those differences are reflected in the price paid for different joints.

The carcase proper consists of the shoulders, forelegs, belly, loin, brisket, back, rump and hindlegs. Other edible portions include the liver, lungs, heart, stomach, parts of the head (brains and cheeks) and parts of the intestine known as chitterlings. The small intestines of sheep and pigs are eaten in the form of sausage skins. The heart, lungs, liver, kidney and oesophagus are often described as 'offals'.

The inedible portions are used in a variety of ways. Bones are used in the manufacture of gelatine and as a phosphatic fertilizer. In the form of a ground, partially defatted powder they are incorporated into animal feedingstuffs. Hides are used in leather and gelatine manufacture, intestines may be used as surgical ligatures, the ductless glands have been used in the manufacture of hormones such as thyroxine, cortisone derivatives and adrenaline. Blood is converted into manure and is sometimes used for the preparation of blood albumin, which finds industrial applications. Small quantities of blood are used in human food in the form of black puddings.

However, the bulk of meat tissue as eaten consists of voluntary muscle, connective tissue and fat. The proportions of these profoundly affect the quality of the meat. A high proportion of connective tissue generally means tough meat, while too high a proportion of fat is regarded by many as greasy and unpalatable. On the other hand, overlean meat is often dscribed as stringy and lacking in flavour. On the balance of these and other factors to be described later depend that elusive characteristic, the eating quality of meat.

The term 'meat' is generally applied to voluntary muscle, connective tissue and adipose tissue. The muscle itself consists of large numbers of muscle fibres, packaged in bundles and surrounded by connective tissue which, to a greater or lesser extent, may be infiltrated by fat cells, referred to descriptively as marbling fat. The connective tissue surrounds the muscle and ramifies through it, providing the soft flexible framework through which the contractile forces in the living muscle are transferred to bone in the form of physical movements. See Fig. 3.1.

Visual inspection of an excised muscle shows it to be surrounded by a sheat of connective tissue known as the **epimysium**. From the inner surface of this, branches of tissue ramify into the muscle separating the muscle fibres into bundles. These separating septa also carry larger blood vessels and nerves and together constitute the **perimysium**. From the perimysium a fine connective tissue framework passes inwards to surround each muscle fibre. This fine framework constitutes the **endomysium**. The size of the muscle fibre bundles influences the texture of the resulting meat, large bundles giving rather a coarse texture. Towards

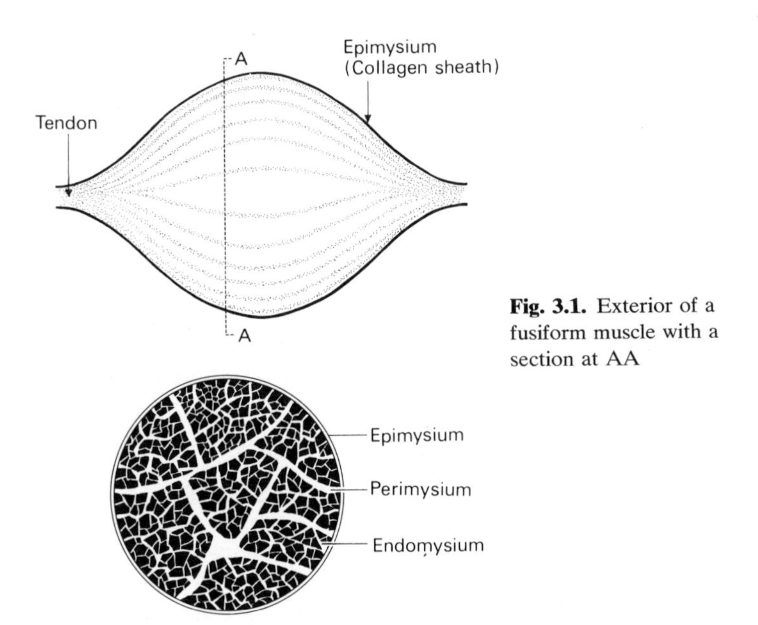

Fig. 3.1. Exterior of a fusiform muscle with a section at AA

the ends of the muscle, the epimysium, perimysium and endomysium blend to form aggregates, emerging as the tendons which attach the muscle to the bone.

The contractile units of the muscle are the fine muscle fibres, which are long, narrow cells containing several or many nuclei, ranging from about 10 μm to 100 μm in diameter (depending on the nutritional plane of the animal). Some of these 'cells' can be of extraordinary length, running up to 30 cm or more in the case of the muscles of larger animals. Surrounding each fibre and underneath the connective tissue of the endomysium is a further sheath known as the sarcolemma. Within this sheath are the **myofibrils** which are bathed in a fluid phase called the sarcoplasm, which, if you like, is just a fancy word for muscle juice! However, this fluid phase is more complex than the term of 'muscle juice' suggests because it contains and supports further bodies. These are (a) the mitochondria or **sarcosomes** which are organised packages of enzymes, (b) **sarcoplasmic lipid bodies** and (c) the **sarcotubular** system of which more later. In addition the sarcoplasm contains other dissolved or suspended substances.

The myofibrils themselves are known to consist of numerous parallel filaments, thick filaments consisting mainly of the protein myosin, and thin filaments consisting mainly of a second protein known as actin, which together provide the contractile mechanism of the muscle.

Adipose tissue is formed by the deposition of fat within the cells of connective tissue. Any connective tissue may become adipose but the greatest tendency is in the muscles supporting the abdominal viscera. The hardness of the fat varies somewhat with the temperature gradient within the body, subcutaneous fat having a lower melting-point than that surrounding the kidneys. For metabolising tissue, it seems that the lipid should be in liquid form and this is exemplified in the extreme by the highly unsaturated nature of fish oils, which remain liquid at sea temperatures approaching the freezing-point of water. The 'greasiness' of a fat at ordinary temperatures is related to its composition. Mutton fat, which is often described as greasy, is a mixture of high-melting-point saturated triglycerides with liquid dioleins. Thus it is oily at ordinary temperatures yet does not completely melt at 42°C.

During fattening of an animal the number of fat cells does not increase but the size of each cell does; increases in diameter of from 10 μm to 100 μm having been observed. In starved animals the fat content of the tissues steadily decreases, the tissues surrounding the heart being the last to be depleted.

The principal fat depots of the body are classified as follows.

1. Subcutaneous: the fatty tissue immediately under the skin.
2. Perinephric: the fatty tissue surrounding the kidney. This tissue is

well-suited to certain cooking purposes because of its high melting-point and is known to butchers and cooks as suet.

3. Intramuscular: the fatty deposits along the line of the perimysium within the muscle and often referred to as marbling fat.

4. Intermuscualar: all other fatty tissue.

So-called lean meat is usually anything but lean. Not only is there usually a network of adipose tissue visible to the naked eye, but the minutest ramifications of the connective tissue carry their quota of fat at the microscopic level. In consequence, the overall composition of every portion of meat on a carcase will vary according to the fattiness of its connective tissue, which in turn reflects the nutritional condition of the animal before slaughter.

The nature of the lipid in the diet of the animal does not greatly affect that of the resulting fat depots in cattle and sheep, although carotenoid pigments carried by the lipid may affect its colour. This is due to breakdown and resynthesis of the lipid during ruminant digestion together with the subsequent absorptive processes. In the case of the pig, however, the lipid deposits and especially the back fat may vary markedly depending on the nature of the fat in the diet. Iodine values ranging from 30 to 110 have been recorded, high values following from diets containing substantial amounts of oil-containing linseed cake or fish meals. Over-soft bacon fat is regarded as undesirable by the trade and by most consumers.

While connective tissue is the vehicle which carries the fat cells, the tissue is not necessarily or invariably infiltrated with fat. On a fat-free basis its principal constituent is the fibrous protein collagen. As present in tissue, **collagen** is tough, insoluble and inelastic, yet is readily converted to water-soluble **gelatine** on boiling. Thus the tougher joints of meat with a relatively high collagen level are used for stewing. While processes such as boiling and stewing soften collagen to the point of solubility by conversion to gelatin, they tend to coagulate and toughen the actin and myosin of the myofibril. Thus fillet steaks, low in collagen are more tender if underdone, while the muscles of the foreleg are more tender when stewed.

Collagen is also abundant in bones and hides and its extraction from these in the form of gelatine and glue forms the basis of a byproduct industry, although natural glue is now largely replaced by synthetic adhesives. Gelatine prepared in this way is widely used in cooking for soups, table jellies, confectionery and the like. The essential amino acid tryptophan is absent from gelatine and in this sense it is of poor nutritional value. However, it is seldom eaten in isolation from other sources of tryptophan (table jelly and milk, for example), and its poor amino acid profile is more a matter of nutritional interest than one of everyday concern. Collagens vary somewhat in their amino acid composition but they are all a little unusual in containing hydroxyamino acids, 4 hydroxy-L-proline being invariably present but 3-hydroxyproline and hydroxylysine are also reported from some collagen sources, and have not so far been found in other proteins.

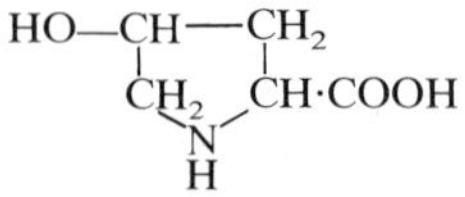

```
HO—CH——CH2
    |    |
   CH2  CH·COOH
     \ /
      N
      H
```

4-Hydroxyproline

As an aside, it is of interest to note that while the complete physiological role of that ubiquitous and much-studied vitamin, ascorbic acid, is obscure, it is known to act as a co-factor in the hydroxylation of proline by the enzyme, proline hydroxylase.

Tropocollagen is the basic structural unit of the native collagen found in muscle. It has a molecular weight of around 300,000 and can be extracted from collagen fibres by dilute salt and acid solutions. In turn, tropocollagen appears to consist of a coil of three polypeptide chains each of a molecular weight of around 100,000. The tropocollagen coils may align themselves in aggregates to form three structurally different types of the collagen fibrils, which ultimately go to make up the collagen connective tissue in the visible sense in which we encounter it in practice.

Connective tissue, however, contains two other constituents **elastin** and **reticulin**. Elastin is practically insoluble, is not softened by cooking, and has other curious if not unique properties. First it is yellow in colour and tissues, particularly the ligaments of the vertebrae, which contain it in quantity, are often referred to as 'yellow connective tissue'. It is only a minor component of skin, tendon, muscle or loose connective tissue. Elastin is resistant to hydrogen-bond-breaking solvents, is heat stable to a temperature of 140°C and resists the action of trypsin, chymotrypsin, pepsin and the cathepsin of kidney or spleen. It is attacked by the plant enzymes ficin and papain. Because of its resistance to mammalian enzymes, elastin is believed to be poorly digested (if at all) by the human gut.

In addition to its obvious presence in the ligaments of the vertebrae, elastin fibres are found in the walls of the large arteries, in the cartilage of ear tissues and in the epiglottis. Cross-linking of proteins is generally achieved in living tissue through disulphide bonds. In elastin, however, cross-links are achieved through two unusual amino acids, desmosine and isodesmosine, and these seem to confer unusual stability on the molecule.

Reticulin is a comparatively minor constituent of connective tissue and has been less exhaustively studied than collagen and elastin, although some consider the collagens and reticulins to constitute a single class of proteins. Morphologically they are distinct, reticulin reacting differently to staining. In particular, it is stained black by ammoniacal silver solutions in contrast to the brown of collagen when treated by the same histochemical stain. In reticulins, the protein moiety has collagen-like properties but it is also associated with bound carbohydrates and fatty acids of which myristic is the most unusual. It does not appear to play an important part in meat quality.

3.3. Meat as a food

The properties of meat as a food depend on more than the structure of the muscle and its associated tissues. To an analytical chemist, the composition of a typical muscle on completion of rigor mortis, and excluding extracellular fat looks as shown in Table 3.1.

A broad analysis of this sort has several intersting features but tells us almost nothing about whether the meat will be good to eat. To understand more we must ask more detailed questions of the individual items listed above.

Water

Three-quarters of the muscle is **water**, which forms the solvent basis of the biochemistry of the muscle mechanism in the living animal and the medium in which chemical changes take place after death. When an animal dies or is slaughtered for food, physical life may be extinguished in a very short time indeed but biochemical processes will continue within the carcase for many hours but at diminishing rates. Most cease, for practical purposes, with the advent of rigor mortis, a process which we shall look at in more detail later. Some of these post-mortem biochemical changes closely parallel those of living tissue. Oxidative phenomena, in

Table 3.1. Composition of a typical muscle

Constituent	%
Water	75.4
Protein	18.0
Fat	3.0
Carbohydrates	0.3
Lactic acid	0.9
Non-protein nitrogen components	1.6
Principal inorganic constituents	0.6
Trace metals and vitamins	0.2

contrast, are inhibited by the absence of circulating blood with its assurance of a continuous oxygen supply. They do not immediately cease but the pathways are halted before completion with important consequences in terms of meat quality. The water is the medium in which this takes place. When these processes eventually cease the water remains, and most of the enzyme systems remain, but they are no longer under physiological control. Some 'random' enzyme action continues. Furthermore the water, with its rich load of nutrients in solution, provides the medium for the microbial growth which is nature's way of disposing of a carcase in the absence of carnivores to do it for her. These microbiological changes are almost all objectionable from a viewpoint of the use of the carcase for human food, and beyond a certain point the meat becomes inedible. If we remove the water from the meat by one or another of the several available processes of dehydration, we arrest biochemical and microbiological change and greatly extend storage life. But in these circumstances, we bring the chemical constituents of the meat into intimate physical contact, and a slower but equally inevitable process of chemical change takes place which ultimately renders the meat inedible.

3.4. Meat proteins

a) The **sarcoplasmic** or muscle juice proteins. These are soluble in water or dilute salt solutions and consist of a complex mixture of over fifty components, many of which are enzymes of the glycolytic cycle, and most of which have now been isolated and crystallised. They are mainly globulins and, while essential to the utilisation of energy in living muscle, some of them may also be of importance in meat in relation to the improvement in tenderness which takes place when meat is stored for one to three weeks at temperatures just above its freezing-point. On thawing frozen meat, sarcoplasmic fluids are lost in the form of an exudate known as 'drip' to the butcher, and this represents waste of both chemical nutrients and commerical value. Much attention has therefore been given to practical ways of minimising such losses, but with limited success. The sarcoplasm also contains the muscle pigment myoglobin together with some haemoglobin.

b) The **myofibrillar** proteins are the working parts of the machinery of muscle and are therefore the most significant of its constituents. They are soluble in concentrated salt solutions, and consist of two main proteins **myosin** and **actin**. Myosin occurs in two major forms, known as heavy and light meromyosins, which together make up about 60 per cent of the myofibrillar material in muscle. Actin also is found as a globular monomer (G-actin) having a molecular weight of around 47,000 and as fibrous actin (F-actin), which is a polymeric form of G-actin, some preparations having apparent molecular weights of over 14,000,000. F-actin appears to take the form of two beaded chains of G-actin monomers twisted into the form of a double-stranded filament, the beads being spherical subunits of G-actin. It is the F-actin which combines with myosin to form the contractile elements in living tissue and the inextensible actomyosin of muscle in rigor mortis.

Thus a picture of muscle emerges as consiting of myofibrils surrounded by the sarcoplasm, which bathes the fibrils in fluid, and packaged together in groups of fibres which are encapsulated by connective tissue. These segments are further grouped into endomysium, and in turn linked through the perimysium to the epimysium, the whole complex unit forming muscle in life and meat after death.

Muscular contraction

Some of the important post-mortem changes affecting meat quality are closely linked with the biochemistry of muscle contraction in the living animal. Energy is obviously required and is obtained from the oxidation of food. Glucose, formed in the liver or absorbed directly from the gut to the bloodstream, is stored in the muscle in the form of glycogen, a glucose polymer with starch-like characteristics. The biochemical steps in this glycolysis process are well known as the Embden-Meyerhof pathway and are outlined in Fig. 3.2.

Adenosine triphosphate (ATP) is the immediate store through which these reactions release oxidative energy to the muscle.

Fig. 3.2. The principal intermediate steps in the Embden-Meyerhof glycolysis pathway showing the links between *in vivo* and *post mortem* situations

Muscle glycogen → glucose-1-phosphate
→ glucose-6-phosphate
→ fructose-6-phosphate
→ fructose-1,6-diphosphate-*ATP*
→ glyceraldeyde-3-phosphate (×2)
→ 1,3-diphosphoglycerate (×2)
→ 3-phosphoglycerate (×2) + 2*ATP*
→ 2-phosphoglycerate (×2)
→ enolphosphopyruvate (×2)
→ pyruvate (×2) + 2*ATP*

Aerobic respiration (e.g. living animal)		Anaerobic glycolysis (e.g. *post mortem*)
$6CO_2 + 6H_2O$	← →	lactic acid (×2)
(through Krebs' cycle)		(catalysed by lactic dehydrogenase)

The substance adenine

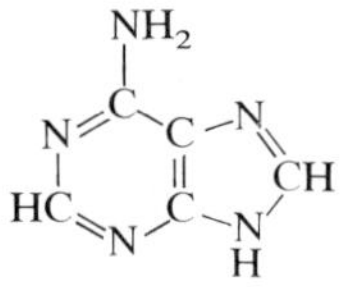

combined with the sugar ribose

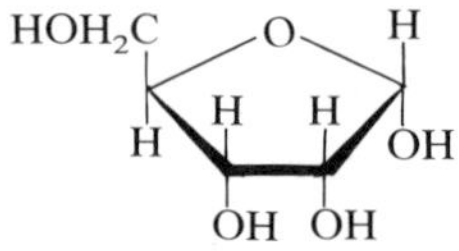

is adenosine and this, in turn, coupled with three phosphate residues

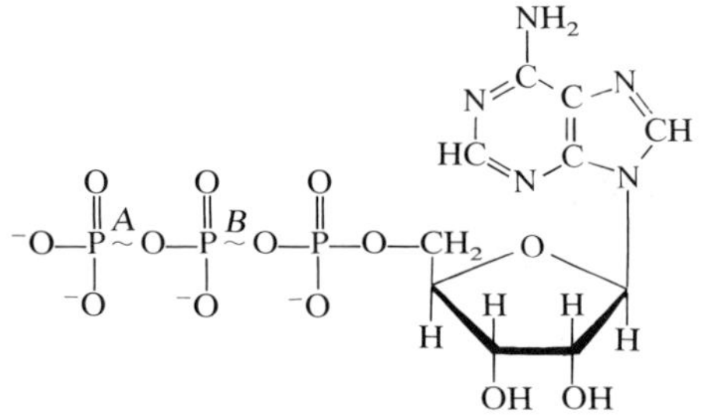

is the substance adenosine triphosphate.

Now ATP is readily converted into adenosine diphosphate (ADP), by hydrolysis at A as shown, with a realease of 31 kJ or 7.4 kcal/mol. In the living animal the ADP produced is ultimately resynthesised to ATP by respiration whereby muscle glycogen as glucose is metabolised through the Embden-Meyerhof pathway to pyruvic or lactic acid. The ~ of the structure shown denotes a bond which releases a high level of free energy on hydrolysis. Under aerobic conditions the pyruvic acid is subsequently fed into the Krebs (or citric acid) cycle emerging at the end of the process as carbon dioxide and water. This breakdown of ATP and its resynthesis provides the link between the chemical energy released by oxidation and the mechanical movement generated in the muscle. The conversion of one molecule of glucose to pyruvic or lactic acid by the Embden-Meyerhof pathway, followed by the subsequent oxidation of these products through the Krebs cycle releases enough energy for the resynthesis of thirty-seven molecules of ATP from ADP.

When muscle is examined under the microscope it is seen to be striated or striped with alternate light and dark bands. The band which appears dark under the ordinary light microscope is anisotropic (i.e. birefringent) under polarised light, and is therefore called the A band. The light band is isotropic and is therefore known as the I band. In the middle of the I band is a dark line known as the Z line. A lighter area in the centre of the A band is known as the H band. In the centre of the H band is a further dark line known as the M line. These optical features provide the structural clues as to the contraction mechanism within a muscle fibre.

The detail of the arrangement can be further examined by the magnifications of up to 200,000 diameters provided by the electron microscope permitting a sight of the arrangements of individual molecules within the system.

The appearance of longitudinal sections of muscle under the light microscope is represented in Fig. 3.3A. From microscopic and crystallographic studies of relaxed and contracted muscle, a preliminary picture of myofibrillar structure emerges as shown in Fig. 3.3B drawn in relation to the original muscle. The thick dark strands are myosin rods and the thin strands attached to the Z lines are actin. There is now general agreement that during contraction the ends of the myosin filaments are drawn towards the Z lines, the actin filaments sliding over them until they meet at the M line and, in extreme contraction, even overlap. The two Z lines in

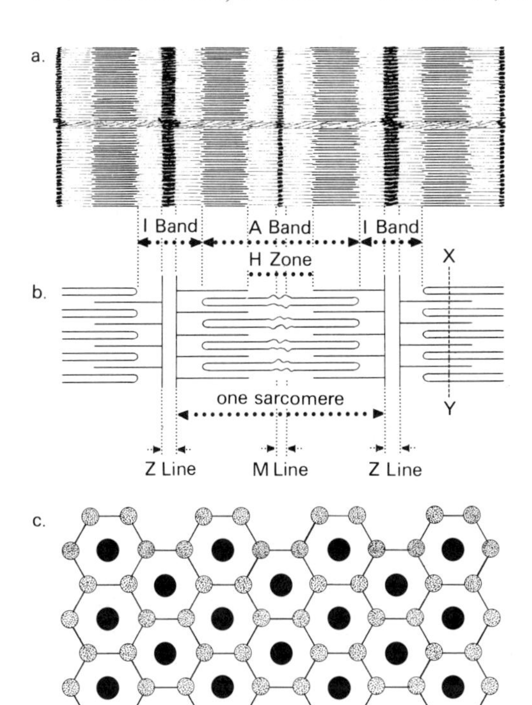

Fig. 3.3. Structure of mammalian muscle. a. As seen under the light microscope. b. A diagram representing the underlying structures. c. A greatly enlarged view of the cross-section XY showing the hexagonal arrangement each myosin rod being surrounded with six actin rods

the sarcomere are thus pulled together, and the action, repeated over a whole muscle, produces the contraction. Relaxation is the reverse process, the myofibrils returning to the resting position shown in Fig. 3.3B.

This planar model can be refined somewhat by examination of a transverse section (at *XY*) under a sufficient magnification in the electron microscope, as shown diagramatically in Fig. 3.3C. The myosin rods are represented as large circles and the actin rods as small circles. Each myosin rod is seen at the centre of a hexagonal array of actin filaments and the interdigitation, described above, represents an interreaction between one myosin rod and its six actin neighbours, each of which is also participating with a similar interaction with the myosin at the centre of the neighbouring hexagon. Again, there is general agreement about this structure, which has been vividly and clearly demonstrated many times by electron microscopy.

However, there is less agreement about the nature of the forces which use the stored energy of ATP to bring about this sliding movement and its subsequent relaxation. A possible scenario of these, reflecting a somewhat simplified form of current thinking on the matter, is as follows.

1. A signal, in the form of an electrical impulse, travels along a nerve fibre and depolarises the sarcolemma or fibre sheath.

2. This brings about changes in the sarcoplasmic reticulum at the Z lines, releasing Ca^{2+}.

3. The calcium ions do two things. First, they release active ATP from its inactive complex with magnesium, allowing the heads of the myosin molecules to link with the nearest actin filament. Secondly, they stimulate the enzyme myosin ATPase.

4. This enzyme splits ATP to ADP thus releasing the energy necessary for the actin filaments (now attached to myosin) to pull or be pulled into the H zone from each side, passing between the myosin filaments.

5. This interdigitation occurs in a series of steps, the muscle shortening overall, as it contracts in response to continuing nerve stimulation.

6. The filaments are locked in the contracted position until nerve stimulation ceases. The tubules of the sarcoplasmic reticulum are able to pump Ca^{2+} out of the system inhibiting ATPase. The ATP level is thus able to rise, as new ATP resynthesised from glycolysis, floods into the system, reforms the magnesium ATP complex and re-establishes the relaxed state.

This explanation of the sequence of events fails to account for the detailed mechanism of the contraction, that is to say, how the actin and myosin filaments slide across one another against the mechanical load applied to the muscle, before they finally lock in the form of actomyosin.

Many years ago, X-ray studies demonstrated that myosin chains could exist either in extended or folded form, which may be rather crudely represented as follows

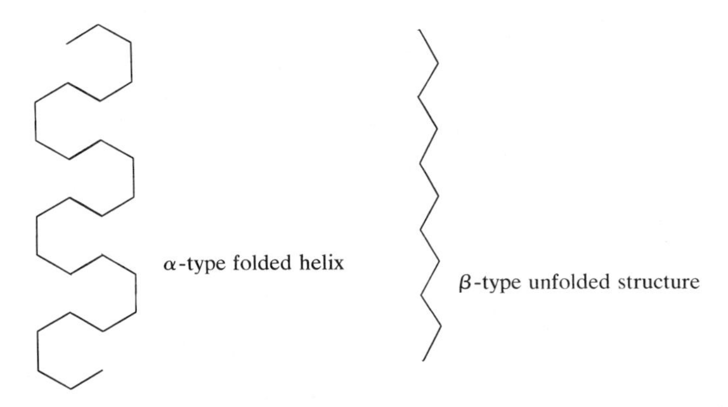

The type of chain formed in any given circumstance will depend on the distribution of charges along the chain, or to put it differently, on the degree of internal hydrogen bonding. This in turn reflects differences in the internal energy levels within the molecule. Furthermore while the two extreme situations are illustrated, intermediate conditions may also exist. In 1963 Davies suggested that transitions between these two myosin structures with their corresponding changes in length may provide an explanation for the sliding movements of actin and myosin filaments in relation to each other.

There are clearly formidable obstacles in verifying this by direct experiment, but in the meantime no new evidence has arisen to invalidate the suggestion. Indeed, increasingly sophisticated structural evidence, which demonstrates the existence of projections or 'heads' at the ends of the myosin molecule which correspond to structural features on the pitch of the G-actin helix, tend to lend strength to the Davies hypothesis. For a variety of reasons which cannot be discussed here, the sliding movement is now believed to be the consequence not of a single transition, but of a sequence of stepwise events occurring between the heads of the myosin molecule and the corresponding structural points on the G-actin helix.

3.5. The conversion of muscle to meat

The muscle of a living animal has a pH of around 7.2 or even slightly higher. After death the pH falls rapidly but in a rather variable way, sometimes to 6.5 and at others to as low as 5.5. It is known that

a) The fall in pH is due to the production of lactic acid in the muscle after death.

b) The extent of the drop varies somewhat from one species to another, and from one muscle to another in a given animal.

c) In general, the greater the drop, the better the keeping quality of the meat.

d) Within a given species there can be great variation in the extent of the drop between one animal and other. With pigs, for example, bone taint, an unpleasant taste in hams, is particularly prevalent in meat of high pH but virtually absent when the pH is really low.

e) The extent of the drop depends on the glycogen reserves of the muscle, the lactic acid being formed from the glycogen by post-mortem glycolysis (see Fig. 3.2) under the anaerobic conditions prevailing in the muscle resulting from cessation of blood circulation. Thus well-fed, well-rested animals with large glycogen reserves will produce meat of lower pH than animals fatigued at the time of slaughter.

f) Associated with this drop in pH is the onset of rigor mortis, in which the muscles of the slaughtered animal lock in the form of actomyosin and are no longer extensible as in the living animal. It is found that, other things being equal, the onset of rigor is generally more rapid in animals of low original glycogen reserves.

g) The full development of rigor mortis normally take place between one and twelve hours after death, but the time of onset varies depending partly on the conditions of slaughter, the nutritional status of the animal, the portion of the carcase concerned and its rate of cooling.

h) The phenomenon of rigor mortis is due to the exhaustion of ATP under the anaerobic conditions of glycolysis prevailing after death. When no ATP remains, no further conversion of glycogen to lactic acid is possible even if residual glycogen remains in the muscle. The depletion of ATP is the final death-point of the muscle.

These considerations bear on the conditions of slaughter in relation to the eating quality of the resulting meat. After the journey to the abattoir, animals should be well rested in a clean lairage with a plentiful water supply and the slaughter system arranged to ensure that the animal arrives in the slaughter-pen in a relaxed and unexcited condition; finally one must ensure that the means used to stun the animal before bleeding the carcase, whether electrical, mechanical (pole-axe, stunning hammer or captive-bolt pistol) or respiratory (carbon dioxide suffocation) is such as to cause the animal the least possible distress.

After stunning, the animals are normally shackled by the hind leg, hoisted into a vertical position and bled by severing the jugular vein and carotid artery in the neck. Thorough bleeding is necessary to minimise subsequent microbiological spoilage, and proper stunning ensures that the heart of the animal continues to beat during the bleeding process thus achieving as complete removal of blood from the carcase as possible. After bleeding, the head, entrails and skin of the animal are removed and the carcase divided into halves to facilitate cooling. During these processes, veterinary inspection of the carcase proper as well as of the removed organs ensures that any diseased animal or any organs infected by parasites are identified and isolated before the meat is released for public sale.

If meat is cooked prior to the onset of rigor it is generally tender. In some tropical countries where spoilage is very rapid this is common practice. For example Chinese cooks in tropical areas such as Malaysia insist on buying what they describe as 'fresh' meat. It is, therefore, local practice to slaughter between midnight and 5 a.m. and to market the still-warm meat in its pre-rigor state between 6 a.m. and 8 a.m. when the traditional Chinese housewife goes to market. The meat is taken straight home and is immediately cooked. Handled thus, it is tender and the very short delay time between killing and cooking makes good sense in terms of minimising the risk of spoilage and of microbial food poisoning under tropical conditions.

When meat proceeds through commercial distribution chains the sale of pre-rigor meat is impossible. Yet meat is tough and unpalatable immediately after it has passed into rigor.

Although meat is subject to microbial spoilage, the degree of which depends on the initial level of contamination coupled with the time and temperature of storage, satisfactory storage for a few days at room temperature or up to about six weeks at temperatures just above the freezing point of meat (−1.5°C) is possible.

This process of 'hanging' or conditioning has long been known to improve tenderness and flavour due to chemical and enzymic changes slowly taking place in the muscle. It is therefore desirable commercial practice to hang meat in carcase form before sale. After three or four days butchers describe a carcase as having passed through rigor or say that the rigor is 'resolved'. To the scientist rigor mortis is the irreversible formation of actomyosin and, if that be accepted, the rigor is never 'resolved'.

Neither scientist nor practical butchers dispute the improvements in flavour and texture resulting from conditioning. The chemistry of the changes involved is as yet only partially understood. Texture is doubtless improved by enzymic breakdown of protein, and flavour is enhanced by the production of inosinic acid and hypoxanthine from the breakdown of ATP, a process which is complete within twenty-four hours of death. The slow breakdown of protein and fat also contribute traces of hydrogen sulphide, ammonia, acetaldehyde, acetone and diacetyl, which are offensive in high concentrations but are equally essential trace constituents in many foodstuffs of highly prized flavour characteristics. However, it will be readily appreciated that if the conditioning process is prolonged, flavour begins to deteriorate.

Meat contains several rather important non-protein nitrogen-containing substances which are believed to contribute to flavour and which are present in quantity in meat extracts. The more important are creatine, carnosine, anserine, inosine and hypoxanthine.

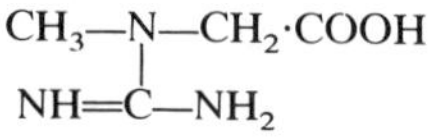

CH_3—N—CH_2·COOH
NH=C—NH_2

Creatine (methylguanidylacetic acid)

Creatine phosphate is involved in one of the pathways through which ATP can be resynthesised from ADP using glycogen as the ultimate energy source. It contributes in this way in the early stages of post-mortem glycolysis. It is present in muscle at levels around 0.5 per cent.

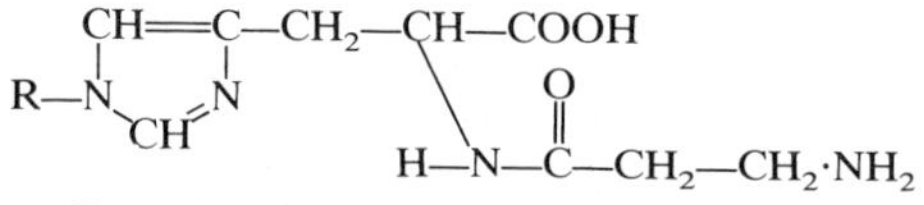

CH=C—CH_2—CH—COOH
R—N   N
CH
H—N—C(=O)—CH_2—CH_2·NH_2

Carnosine (R = H) and anserine (R = CH_3)

Carnosine (β-analylhistidine) and anserine β-analylmethylhistidine appear to act as buffers in living muscle helping to regulate pH. The quantities found average around 0.3 per cent but vary between different muscles as also between different species. It is noticeable that they are found at high levels in muscles which may be deprived of oxygen, as shown below

Whale muscle (anaerobic):	high anserine, low carnosine
Rabbit (average anaerobic):	moderate anserine, low carnosine
Pigeon breast (aerobic):	low anserine, low carnosine
Heart muscle (very aerobic):	both absent

In oxygen-restricted muscles, lactic acid tends to build up and the presence of buffering substances is then essential.

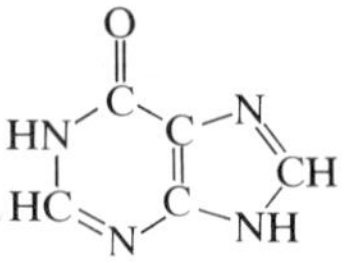

O
HN—C—C—N
HC=N—C—NH
CH

Hypoxanthine (6-oxypurine)

Hypoxanthine is a breakdown product of ATP as follows

$$\begin{aligned} ATP &\rightarrow ADP + H_3PO_4 \\ &\rightarrow \text{adenylic acid} + H_3PO_4 \\ &\rightarrow \text{inosinic acid} + NH_3 \\ &\rightarrow \text{inosine} + H_3PO_4 \\ &\rightarrow \text{hypoxanthine} + \text{ribose} \end{aligned}$$

Ammonia, urea and amino acids are also present in meat in small quantities. There is now some evidence linking the presence of these and other substances with production of cooked meat flavours.

3.6. Vitamins and minerals

As might be expected, meat provides useful amounts of the B-group vitamins and vitamin A. The latter is present in some muscular tissues at low levels, in kidney at reasonable levels and abundantly in liver, which may contain as much as from 1,200 μg to 13,500 μg retinol equivalents/100 g as against an adult daily recommended intake of 750 μg. While only present in trace quantities in muscular tissue, vitamin D is found in beef liver at levels of around 1 μg/100 g, 10 μg per day being sufficient to protect a child against rickets.

While vitamin C is virtually absent from muscular tissue, it is found in useful amounts in liver and kidney (15–30 mg/100 g) and this may provide an explantion of reports of subjects remaining free from scurvy on an all-meat diet. The Norwegian explorers, Nansen and Johanson, after leaving their ship the *Fram*, spent nine months including the winter of 1895–96 on Frederick Jackson Island and remained free from scurvy despite having no access to fruit or vegetables and living mainly on fresh walrus or bear meat. In 1900 the travellers Jackson and Harley described how six Russian priests arrived in the Yugor straits in the autumn accompanied by a small Russian boy. The priests were prevented by their vows from eating meat and had to subsist on salt fish. The boy ate fresh reindeer meat. When spring came, he was the sole survivor, the other having died of scurvy.

In western diet meats are regarded more as an important source of B-group vitamins. They contribute around 0.1 mg thiamin, 0.2 mg riboflavin and 5 mg nicotinic acid/100 g. It must also be remembered that they contribute in a unique way to the vitamin B_{12} intake in human diet.

Meats are generally rich in iron and phosphorus but contain only small quantities of calcium. In addition they contribute useful amounts of potassium magnesium to human diet. Liver is particularly rich in iron, containing about 10–20 mg/100 g.

3.7. Meat pigments

Meat fat contains varying amounts of carotenoid pigments derived from the animal's food. As the beast grows older the fat darkens in colour from creamy white through cream to yellow and this provides the butcher a rough index of the animal's age. The types and quantities of pigment present can vary very considerably.

Muscle pigment is principally myoglobin along with much smaller quantities of haemoglobin. Changes in meat colour are obvious to even the most casual observer of a butcher's shop, and at the point of purchase bright clear red meat is more attractive than a dull red with purplish tones. When cooked, meat turns grey or brown unless it has been previously treated with a meat-curing brine (normally containing salt, nitrate and nitrite) when the cooked colour is an attractive pink. Myoglobin and haemoglobin change colour readily depending on their state of oxidation. Some of these changes are worth remembering.

The pigments themselves consist of porphyrin nucleus (often referred to as the haematin residue), attached to a protein known as globin. The porphyrin residue, in isolation known as haemin, is of the following form

Pyrrole

Haemin

In the living tissue the iron is in the ferrous state and the molecule has the property of forming a loosely associated complex with oxygen, haemoglobin acting as the blood carrier of respiratory oxygen. However, in meat the iron may be in either the ferrous or the ferric state depending on the treatment the meat has received. Furthermore, the globin may be denatured (for example by the effect of heat in cooking) or in its native state. Additionally, the nitrite used in meat curing also reacts with myoglobin, giving rise to further chemical changes depending again on its concentration and on subsequent heat treatment of the meat. A number of other substances, such as hydrogen sulphide, also react with myoglobin

and their products are occasionally found in meat, but those are less important for our present purposes.

Thus, myoglobin is purple-red in colour while oxymyoglobin is bright red, in both cases the globin being in the form of the native protein. However, oxidation of myoglobin (as opposed to oxygenation) with the conversion of the iron atom from the ferrous to ferric state, produces the brown compound, metmyoglobin, again with the globin in the native state. The effect of heat is to denature the globin and often to detach it from the haematin residue. The colour of the resulting compounds, myohaemochromogen or myohaemichromogen, will be red or grey-brown respectively depending on whether the iron remains in the ferrous or ferric state.

When meat is treated with curing solutions containing nitrite, nitric oxide myoglobin (sometimes called nitrosomyoglobin) is formed. The iron remains in the ferrous state and the colour of the meat become bright red or pink. When such meat is cooked the compound nitric oxide haemochromogen is usually formed, which also has a bright pinkish red colour. Thus well-cooked beef is grey brown while fried bacon remains red.

3.8. Bacterial spoilage of meat

The muscular tissues of a healthy animal are sterile. However, the process of bleeding introduces bacteria from the skin of the animal and the knife of the operative into the bloodstream. These are quickly carried throughout the tissues by the residual circulation. The situation is aggravated by some invasion of the bloodstream by intestinal bacteria during slaughter.

External contamination of the meat from the environment begins in the abattoir when, despite the most stringent precautions, the presence of animal hides with their adhering soil and gastro-intestinal contents together with airborne microorganisms, the necessary use of knives, cleavers and saws and even the water used for cleaning the carcases add to the loading produced from the environment. The staff employed inevitably add to the contamination situation. All of these sources can be controlled, but none eliminated, by attention to hygiene.

Animal carcases will support the aerobic growth of moulds and yeasts and the aerobic or anaerobic growth of bacteria. In practice, growth depends in part on the initial contamination, and in part on the storage conditions. The more usual effect is decomposition and putrefaction brought about by non-pathogens. Since it is normal practice to store meat at temperatures of 5°C or less, the most common type of growth is from psychrophilic groups such as *Achromobacter, Micrococci, Lactobacilli* and *Pseudomonas.* Mould growth is not uncommon on the surfaces of chilled meat, *Penicillium* and *Cladosporium* species tending to be prominent amongst the wide range of possibilities.

The principal factor affecting growth is, of course, temperature. If the humidity in the store is low, surface drying of the meat takes place and this in turn both delays development of microorganisms and influences the selection of species which find a sufficiently favourable environment for growth. The presence or absence of nitrite and salt influences both growth rates and the types or organisms developing, nitrite being of particular value as an inhibitor of *Clostridium botulinum*. The effect of pH has already been mentioned, the more acid the meat, the more restricted is the range of organisms which can grow. There may also be development of food poisoning organisms such as the *Salmonella* species, *Cl. welchii. Staphylococci* and *Enterococci* and even *Cl. botulinum*, although fortunately poisoning by this most deadly of organisms is rare, compared to outbreaks associated with the other species.

The storage of fresh meat and the progress of conditioning produce the best quality product by a series of compromises. Low pH, low temperature and low humidity result in the maximum suppression of microbial activity in fresh sides of meat. But these conditions will normally produce excessive drying of the meat surfaces with loss of 'bloom' (the trade term applied to an attractive appearance) as well as excessive weight loss, by evaporation, which also represents money loss to the trader.

At rather higher temperature and humidities (say around 4°C and 90 per cent relative humidity), the conditioning process proceeds more rapidly but some surface mould growth may ensue. Temperatures of 4–6°C are also used for meat curing, but here the growth of halophilic organisms in the curing brines may be deliberately encouraged because of their ability to reduce nitrate to nitrite. After curing, sides of bacon are normally stacked in cellars held at 4–6°C for seven to fourteen days to allow the curing salts to diffuse evenly through the tissues. During this time the characteristic bacon flavour develops but little is known about the reactions responsible for this, and whether they are of microbiological origin.

3.9. Faults in meat

Toughness is the fault most frequently encountered in complaints of meat quality. Apart from the anatomical and environmental factors already discussed, toughness and tenderness are related also to the conditions under which meat is handled after slaughter. Perhaps the most important single factor underlying the development of toughness is the post-mortem contraction of muscles in the course of rigor mortis. In general, excessive contraction leads to toughness. There are two levels of temperature which encourage this. If excised muscle is held at body heat, without any attempt at cooling, post-mortem glycolysis is rapid with associated excessive shortening and consequent toughness. As muscle temperature is reduced from around 40–15°C the severity of the toughening diminishes. At still lower temperatures, however, the rate of contraction increases and again toughness is the outcome. This latter and imperfectly understood phenomenon is known as cold shortening.

The commercial practice of suspending sides of beef by the rear leg after slaughter has the effect of minimising shortening, most of the muscles being extended by the weight of the carcase, although some have a limited degree of freedom to contract. The rate of heat loss from a large beef carcase is slow, and the temperature fall in different parts of the carcase will be non-uniform as the carcase cools. Thus cold shortening and consequent toughness is less likely to affect muscles deep in the carcase, such as the rump, post-mortem glycolysis having proceeded to completion long before refrigeration can lower the temperature to below 15°C.

Very rapid post-mortem glycolysis is observed in pig muscle from time to time and leads to what is known as a white muscle condition or sometimes as PSE (pale soft exudate) of muscle. The cut muscle has a whitish appearance and exudes enough fluid due to reduction of its water-holding capacity to be described as 'wasseriges Fleisch' or 'watery meat' by German food technologists. The fault is associated with high slaughter-house temperatures and hot weather together with violent struggling of the animal immediately before slaughter. It appears that a rapid fall in pH after death leads to denaturation of the contractile proteins with a corresponding loss of water-holding capacity. However, other factors are involved, as reflected in the fact that the condition is associated with overt stress in the affected animal and that incidence varies with the breed of pig. Some experiments have suggested that the pale colour is due to denaturation of myoglobin. Others indicate that myoglobin levels are unusually low in muscles so affected. The condition is of some economic importance, but despite considerable research efforts it is still not fully understood.

Meat of bright appearance and a clear red colour looks more attractive to the purchaser than dark dull-coloured meat which may often have an unpleasant purplish tinge. This is known as dark-cutting beef and is associated with cattle rather than with pork or mutton. It is common in animals of excitable temperament, especially if they have been subjected to stress (such as a train journey) before slaughter. Unlike PSE meat, dark cutting beef is of high pH and its occurrence is associated with depletion of glycogen reserves due to muscular short-range tensions (not necessarily manifest in the form of external movement) prior to slaughter.

These patterns of change can be summarised as follows:

1. A slow steady decrease of pH to around 5.7 or lower yields normal meat.

2. A very rapid decrease of pH to 5.4 or even 5.1 in 0.5–1.5 hours produces the PSE effect in pork muscle.

3. A slow decrease in pH to ultimate values of 6.0 or higher yields the dark firm dry muscle of dark cutting beef. At these high pH values the water-holding capacity of the muscle is enhanced giving the dry appearance.

Further reference

Meat Science (1979) Lawrie, R. A., 3rd edn, Pergamon, Oxford.

The Science of Meat and Meat Products (1971) Price, J. F. and Schweigert, B. S. (eds), 2nd edn, Freeman, San Francisco.

The Physiology and Biochemistry of Muscle as a Food 2 (1970) Briskey, E. J., Cassens, R. G. and Marsh, B. B., University of Wisconsin Press, Madison.

Muscle (1972) Smith, D. S., Academic Press, London.

Machina Carnis (1971) Needham, D. M., Cambridge University Press, Cambridge.

Meat and Meat Products, Williams, E. F., in *Quality Control in the Food Industry* (1968) Herschdoerfer, S. M. (ed.), vol. 2, pp. 251–301, Academic Press, London.

Meat, Paul, P. C., in *Food Theory and Applications* (1972) Paul, P. C. and Palmer, H. H. (eds), pp. 335–494, Wiley, New York.

Animal Foods, in *Recent Advances in Food Science* (1962) Hawthorn, J. and Leitch, J. M. (eds), vol. 1, pp. 49–112, Butterworths, London.

The Principles of Meat Science (1975) Forrest, J. C., Aberle, E. D., Hedrick, H. B., Judge, M. D. and Merkel, R. A., Freeman, San Francisco.

4

Fish

4.1. Introduction

Since both the structure and contraction mechanism of fish muscle are similar to that of meat, it is convenient to consider the topics sequentially. One difference of outstanding importance, however, is the fact that while the bulk of meat supply comes from domesticated animals and only a small fraction is hunted wild, the reverse is true of fish. Production of marine fish by farming is trivial, although freshwater fish are now produced on a modest but significant scale.

The world's oceans cover 71 per cent of its surface and are sometimes assumed to contain a huge reservoir of protein as yet untapped. In fact, most of the oceans are comparatively barren, the great commercial fisheries being located in the relatively shallow waters of continental shelves and elsewhere, and are of limited extent. The exploitation of marine resource by conventional means may now be reaching its peak, as Table 4.1, showing world catches, indicates.

The United Kingdom, once described as an island of coal surrounded by fish, has steadily invested in larger, faster, more powerful and better-equipped fishing boats, yet her yields of fish are only being sustained with difficulty (see Table 4.2).

However, the British situation is by no means as simple as the figures indicate. For example very roughly half the weight of fish caught is bones, skin, fins, head, tail and intestines. From these offals of our catch of around 1,000,000 tonnes supplemented by catches of fish not directly wanted for human consumption, we make about 100,000 tonnes of fish-meal. We import a varying amount of fish-meal from abroad but in a

Table 4.1. The world catch of marine and freshwater fish
The figures are based on the U.N. statistical year-books and the F.A.O. *Year-Book of Fisheries Statistics.* On a world-wide basis estimates necessarily involve errors but the present trend of the figures suggests that an asymptote is now being reached

Year	World catch
	million tonnes
1958	39
1968	64
1972	66
1973	67
1974	71
1975	70
1976	74

Table 4.2. U.K. catch of marine and freshwater fish
Figures for U.K. landings suggest to some observers that less intensive and less expensive fishing methods may, in the long term, produce higher landings. In the short term the immediate livelihood of large numbers of fishermen depends on sustaining yields even if fishing costs rise steeply. (As in Table 4.1 the figures have been rounded off)

Year	U.K. catch
	million tonnes
1971	1.11
1972	1.08
1973	1.13
1974	1.09
1975	0.98

typical year it may amount to 500,000 tonnes. This total of 600,000 tonnes is used as animal food and we therefore consume it indirectly in the form of pork, bacon, eggs and chicken. Since about 5 tonnes of fish or fish offals are required to make 1 tonne of fish meal, the U.K. uses the equivalent of 3,000,000 tonnes of fish in this indirect form. The U.K. population is about one-seventieth of the world population but we use about one-twentieth of the world's fish supply. British demand for fish is thus high, as is its sensitivity to problems of overfishing, with its concomitant depletion of fish stocks.

4.2. Fish food chains

The existence of sea fisheries depends on photosynthesis in surface waters (known as the photic layers) by floating diatoms. These are microscopic unicellular algae, which, together with dinoflagellates and a few other algae, provide the primary food source for all life in the sea. Their growth depends on sunlight and on the presence of the necessary inorganic salts in the water. These diatoms supply the food of a number of larger creatures, such as larval fish swimming in the surface waters and molluscs such as oysters, mussels and cockles.

The term **plankton** is used to describe any more or less passively floating or drifting minute plant or animal of sea or lake. Plankton capable of photosynthesis are often referred to as producer plankton and sometimes as phytoplankton. They are of small size (2–50 μm). The larval forms which make phytoplankton their food are known as zooplankton and range in size from 0.5 mm to 50 mm. Phytoplankton tend to develop in cycles or flowerings when they may become so numerous as to colour the water. The flowerings usually appear in late spring or early summer and again in the autumn, and disappear after a few weeks. The arrival of these flowerings is due to the presence of suitable salts in the water. These salts are only required in small amounts, together with sunlight which penetrates to a depth of 20–100 m (the photic zone) depending on latitude and water turbidity. When the critical nutrient (usually nitrate or phosphate in quantities around 30–200 mg/m^3) is exhausted, the flowering ceases and is renewed again when mixing of deeper and surface water occurs due to thermal inversion. The nutrient salt concentrations in the deeper waters are continuously renewed by slow decomposition of biological materials of all kinds on the sea bed.

The zooplankton are small swimming animals feeding on phytoplankton. They include the larvae of the principal food fish whose first food after exhaustion of the yolk sac consists of diatoms. In their turn the zooplankton form the food of post-larval fishes and certain adult food fishes (such as herring, mackerel and sprats) and an immense number of other sea animals, some of which are the chief foods of adult food fishes.

The zooplankton have a life of from 2 to 10 months and form a more or less continuous link between phytoplankton and larger fish. They can thus be regarded as a bank or store spreading the energy of the short but frantic bursts of diatom activity over a longer period.

In contrast to land animals, where the species used for meat are herbivores, the species of fish used for food are carnivores. Their feeding habits are complex but are confined to waters seldom deeper than 200 m, such as the continental shelves, the North Sea Banks, and the Newfoundland Banks, where plankton production is much greater than over the deep waters of the oceans.

The chief method of harvesting is to catch the bottom-living fish by means of a trawl, which is simply a large netting bag dragged over the sea-bed by a sufficiently powerful ship. The fish caught in this way include cod, haddock, whiting and plaice. Bottom-feeding fish are referred to as demersal. Pelagic fish, such as herring, mackerel, pilchards, sprats, anchovies and sardines, are free-swimming and generally feed on zooplankton in surface waters. They are caught by curtains of netting known as ring or drift nets whose tops are supported by buoys.

White fish are normally gutted at sea and packed in boxes interlayered with ice for landing. Of the common white fish (cod, haddock etc.) between 44 and 52 per cent of the landed (i.e. gutted) weight forms the edible portion. Pelagic fish, such as herring, are landed ungutted and between 50 and 55 per cent of the ungutted weight is edible. If the skin is removed a variable deduction of 2–5 per cent is necessary depending on the relative size of the fish. Between 50 and 60 per cent of flat fish is edible, crustaceans yielding around 40 per cent and shellfish ranging from about 12 to 30 per cent. These figures must be kept in mind when comparing fish prices at the quay with those in the shops.

4.3. The chemical composition of fish

Fish flesh contains no carbohydrate except glycogen. Their connective tissue is less obvious and more diffuse than that of meat animals. It is useful to consider first the fat content of their flesh and for this purpose to divide them into six groups.

1. Round fish (cod, haddock, whiting, saithe, hake etc.) have flesh of low fat content, ranging from 0.1 to 0.9 per cent.
2. Flat fish (plaice, flounders, lemon sole) have flesh of fat content normally between 0.3 and 4 per cent, the actual figure varying with the season. With halibut, the range is greater, figures from 0.3 to 9.0 per cent having been recorded.
3. Pelagic fish (herring, pilchards etc.) are very fatty when in prime condition, figures of up to 22 per cent in herring flesh having been recorded. After spawning the figure drops markedly and may fall to as low as 1 per cent. As fat is lost it is replaced by water so, water being denser than fat, the fish may gain weight while 'slimming'. In pelagic fish that fat is not stored in depots as with animals but is distributed through the tissues in a diffuse fashion.
4. *Salmo* species (salmon, sea trout and brown trout) are intermediate in fat content between white fish and pelagic fish, and again the actual levels found relate to the onset and completion of spawning. Recorded levels in trout range from 1 to 10 per cent while salmon flesh in fresh-run fish produces figures as high as 14 per cent, but this declines in fresh water and may drop below 1 per cent after spawning.
5. Shellfish (oyster, mussel, scallop, winkles and cockles) as a group have a low fat content of from 0.4 to 2.5 per cent.
6. Crustaceans (crab, lobster, prawn, and shrimp) are also generally low in fat, the usual figures being similar to those of shell fish, although crab meat sometimes contains up to 6 per cent.

Fish lipids must remain liquid at sea-water temperatures. Characteristically therefore they are highly unsaturated. Triglycerides serve as a reserve source of energy and this depot fat is stored in different anatomical locations. In the cod, for example, muscular tissue contains around 1 per cent or less of total lipid and of this, only about 3 per cent is triglyceride. On the other hand cod liver is a rich source and contains virtually all the depot fat. In many flat fish, such as halibut and sole, the depot fat is principally located just below the skin, and the muscular tissue proper contains similar levels of fat to that of cod. The non-triglyceride lipid varies little with the nutritional status of the fish. In general it consists of lecithins and other phosphatides, some waxes and alcohols, cholesterol and cholesterol esters and small quantities of free fatty acids.

4.4. Water content of fish

The water content of a large range of the most common low-fat species of fish used for human food falls within the range 78–81 per cent and even pelagic fish fall into this group when their fat content drops to very low figures at the end of the spawning season. Turbot, halibut, mullet, bass and dogfish have a moisture content range around 75–78 per cent but these usually contain around 3–5 per cent fat. Indeed looking at lists of analysis one is impressed by the fact that moisture plus fat often adds up to around 78–80 per cent. Protein contents (nitrogen ×6.25) tend to be around 18 per cent in many species, thus leaving 2 per cent for other constituents such as water soluble substances (extractives), vitamins and minerals. Generalisations can sometimes be misleading, but this general pattern is a useful rough guide providing it is treated simply as such.

The most important exceptions are salmon and eel species, both having a lower water content (around 67 per cent) and a higher fat content (around 14 per cent). The energy value of both of these species is higher per unit weight served than most other fish. Perhaps this is the reason for both of these being described as 'filling' foods, meaning that people feel satisfied with smaller portions of these than they sometimes do with other fish. Crustaceans are also exceptional in having lower water contents (around 70 per cent) and higher protein contents (around 20–22 per cent) than most other marine species used for human food.

4.5. Glycogen and lactic acid in fish muscle

The course of development of lactic acid after death is the same in fish as in meat. However, the ultimate pH reached is usually higher for three reasons. Fish muscle appears to be designed for short bursts of high activity rather than sustained and continuous effort. The glycogen reserves are low. Fish muscle also contains the substance trimethylamine oxide, which has a strong buffering action and thus tends to oppose a pH drop. These factors combine to set the limit of the observed drop to about pH 6.2 for well-rested specimens. However, commercial methods of catching by net involve the fish in a prolonged struggle, and the glycogen may be almost completely depleted before death. As a result the pH of commercially caught fish flesh is frequently around 7.0, which is close to the living value of 7.3. This is clearly an important factor in the storage life of fish. Others will be considered later.

4.6. Fish muscle composition

The fine structure of fish muscle is similar to that of meat and the development of rigor mortis is essentially the same, except when modified by the circumstances described in the previous paragraph. The main constituents of muscle are protein and water with varying amounts of fat as described previously.

Like other vertebrates, fish have smooth, cardiac and skeletal muscle, the skeletal muscle being of principal interest as a source of human food. Since fish move by alternate contractions of the muscle mass on each side of the vertebrae, the muscles are arranged in segments known as myotomes. These are separated by layers of connective tissue or myocomma. Fish muscle cells span these layers of connective tissue in which their ends are embedded, there being one myotome for each vertebra. During cooking the myocomma degrade to gelatine with the separation of the muscle tissue into flakes. Fish connective tissue differs from that of meat in that it consists entirely of collagen, elastin and reticulin being absent.

Under certain conditions of catching, fish may go into rigor mortis at comparatively high temperatures. Especially above 17°C the associated contractions can be so strong that the connective tissue is torn apart and 'gaping', or separation of the tissue flakes takes place, with a consequent diminution in commercial value.

Fish muscle proteins are nutritionally as good as meat proteins and their biological value is little affected by normal processing operations such as freezing and drying. Even fish protein concentrate, made by treating comminuted fish with an organic solvent to remove water and lipid simultaneously, retains a high biological value after final drying despite the chemical insult of the extraction method.

On the other hand, processing operations do affect the eating properties of fish, which may develop a tough rubbery texture and may also become discoloured. The severity of these effects vary from species to species.

Fish muscle proteins fall into four groups.

a) The water-soluble proteins are mainly sarcoplasmic in origin (i.e. from fish muscle cell juice) and amount to about 20 per cent of the total protein content of the muscle. Over fifty different enzymes have been identified from this fraction and it is therefore of some importance in relation to flavour changes on storage. On the other hand, it appears to play only a minor role in determining fish texture.

b) The salt-soluble proteins consist mainly of myofibrils and therefore of actin and myosin. Post-mortem changes in these and other changes brought about by processing, especially by freezing, play a dominant role in determining fish texture. Much has been written about the mechanisms involved but finality of conclusion is still awaited. Cross-linking reactions appear to take place in a variety of ways resulting in aggregates of high molecular weight. Furthermore, the hydration shells of the proteins are altered by processing conditions (especially freezing) which, in turn, affect their mechanical properties. This group of proteins comprises about 75 per cent of the total protein in the muscle.

c) The insoluble proteins (or stroma) consist chiefly of connective tissue and cell walls and make up around 5–10 per cent of the total protein in most commercial species.

d) The pigmented or chromoproteins include the usual haemoglobin and myoglobin of blood and muscle as well as the cytochromes. During freezing of whole fish at sea there may be leaching of these into white fish muscle with subsequent patches of discolouration which reduce the commercial value of the product.

4.7. Vitamins and minerals in fish

All fatty fish contain vitamins A and D in the flesh but white fish has very little of these except in the liver. Fish livers furnish the richest natural sources of vitamin A, halibut liver being quite remarkable in its content of this substance, figures of up to 100,00 μg retinol/g halibut liver oil having been recorded. Other fish liver oils, such as those of cod and shark, are also excellent sources of the vitamin but at much lower levels of around 2,000 to 12,000 μg of retinol/g. Fish liver oils are also excellent sources of vitamin D, halibut again being the richest of these with levels of around 50–100 μg/g oil compared with cod liver oil levels of around 20–25 μg/g.

For practical purposes fish muscle is devoid of vitamin C, although enough is present in fresh fish roe and liver to provide protection against scurvy for peoples such as the Eskimos whose diet is mainly of fish and meat. Fish muscle is a reasonably good source of B-group vitamins. For example, thiamin requirements are linked with total energy expenditure. Fish muscle contains around 0.05 mg thiamin/100 g tissue, and compared with whole wheat (0.4 mg/100 g) the figure is not impressive. However, converted to an energy basis the levels are closer, fish muscle being around 0.7 mg/1000 kcal (0.17 mg/MJ) while wheat is 1.2 mg/1000 kcal (0.3 mg/MJ). Fish is also a good average source of other B vitamins, typical figures being 4 mg/100 g for nicotinic acid, 0.3 mg/100 g for riboflavin, and 50 μg/100 g for folic acid.

As a source of minerals, fish tissue contributes usefully to intakes of calcium, magnesium and phosphorus, herring, shellfish and some crustaceans being particularly good calcium sources at levels around 100–200 mg/100 g. Fish contains less iron than lean meat as would be expected from the lower levels of blood, about the same amount of copper and, in the case of marine fish, about one hundred times the level of iodine. The contribution to iodine intake of marine fish is of particular significance in British diet, since the only rich source of iodine is seafood. In 1964, for example, Sir Edward Wayne obtained the following figures for fish in Glasgow: haddock 6590 μg/kg, whiting 650–3610 μg/kg and herring 210–270 μg/kg. He estimated that diets in Glasgow provide from 40 μg to 1000 μg per day, and that if sea-fish is taken for one or two meals per week, the average daily intake will be well over the 150 μg/day recommended by the British Department of Health and Social Security as the minimum requirement.

4.8. Extractives and nitrogeneous bases

In addition to the nitrogeneous bases found in meat (carnosine, anserine and creatine) fish contain urea and trimethylamine oxide $(CH_3)_3NO$, whose presence is characteristic of marine species. It is believed to function in the osmotic regulation of fish, and can act as a hydrogen acceptor, being reduced by glutathione. On heating to a high temperature, trimethylamine oxide is degraded to a mixture of dimethylamine and formaldehyde. As a result canned fish may contain appreciable quantities of formaldehyde, a consequence which has given some trouble from time to time to regulatory authorities, since formaldehyde is a forbidden preservative in most countries. Gadoid species (cod, etc.) contain an endogeneous triamine oxidase system which converts trimethylamine oxide to dimethylamine and formaldehyde. The reaction is of significance in frozen fish.

It has been claimed that trimethylamine oxide and the betaines (trimethylammonium bases) are largely responsible for the off-flavours which develop in fish on storage. Trimethylamine oxide can certainly be degraded by bacterial action to tri- and dimethylamine, which do smell like decayed fish. However, staleness and deterioration of fish is more complex than this simple explanation permits and there is not a very good correlation between trimethylamine content and organoleptic data during the process of fish spoilage.

4.9. Fish spoilage

Generally speaking, the flesh and body fluids of newly caught healthy fish are sterile. Nevertheless the skin slime, the gills and, in feeding fish, the intestines carry heavy bacterial loads. These range from 10^2 to $10^5/cm^2$ of skin, or from 10^3 to $10^7/cm^2$ of gill tissue for fish caught in temperate zone waters. The gut contents would be of the order of 10^3–10^8/g. In tropical waters the figures would be higher by two or three orders of magnitude. In cooler waters psychrophiles are more abundant and mesophiles represent only about 5 per cent of the total flora, while in tropical waters the latter make up about 55 per cent of the total numbers. Thus it would be expected that, in general, storage in ice would have more effect in retarding spoilage in tropical fish than in fish from the temperate zones. This has been found to be the case both by experiment and in commercial practice. For example, North Sea fish are unacceptable after nine to twelve days storage in ice, while Indian-caught species required up to forty-five days in ice to reach a similar degree of staleness. However, the icing of fish in tropical water is rare, despite the high rate of spoilage at ambient temperatures. One hour's exposure to tropical temperatures is said to equate roughly in staling to one day on ice. Thus, if icing is used in the tropics it must be done immediately after catching. Consumers in the tropics will accept fish at staleness levels unacceptable in temperate climates.

In general, the bacterial flora found in the slime covering fish skins reflects the flora of their environment, and this has been confirmed by examining the flora of an Atlantic salmon in the sea with that of the same salmon after it had entered fresh water to spawn. Fish from polluted waters will therefore carry the polluting organisms on their skins, gills and perhaps intestines. Such fish have frequently been incriminated in food poisoning outbreaks. Of special significance is the occurrence of *Clostridium botulinum* of the psychrophilic types E and F in certain waters, particularly those of the Baltic and the great lakes of North America. Fortunately this organism is rarely observed in British coastal waters. It has recently also been found in small numbers in farmed freshwater fish such as trout. *Salmonella* are also found, as would be expected in polluted waters. However, the predominating flora in fish from temperate water are non-pathogens and include the genera *Pseudomonas, Achromobacter, Vibrio, Flavobacterium* and *Coryneforms.*

Fish slime and fish flesh provide an excellent bacteriological medium. Fish coming aboard a trawler may lie on deck in the sun for several hours before being gutted, washed and stored in crushed ice. Moreover, on a pitching vessel at sea it is not possible to prevent the gutted fish from coming into contact with the discarded guts. The decks of the ship and the boots, hand and clothes of the fishermen all add to the general contamination of the environment. It is impossible to ensure that the ice used does not contribute its own quota of flora to the situation. As a result of this range of adverse factors, spoilage begins rather quickly despite the ice, and iced fish in temperate waters become stale in ten to twelve days. If the steaming time of the catching vessel from the fishing grounds to port is, say, four days, and if it takes seven days fishing to fill the holds, the first fish caught will have a total count of about 10^8 micro-organisms/cm^2 by the time of landing and even if used immediately will be stale and only marginally acceptable as 'fresh' fish.

This level of contamination is not visible and obvious except to the experience eye, and much thought and experiment has gone into attempts to produce rapid laboratory measures of the development of staleness. None of these can as yet be regarded as entirely satisfactory in all circumstances, but the dielectric 'freshness' meter designed by Jason and Lees of the Torry Research Station, Scotland, appears to offer as good an approach to this problem as has yet been made for fresh fish. The instrument consists of two electrodes which when placed on the skin of the fish adjacent to the lateral line give a measure of the Q factor of the tissue, the Q factor being defined by the expression

$$Q = 2\pi fCR$$

where C is the capacitance, R the resistance and f the frequency of measurement. In practice, the frequency used is 2 kHz. The readout correlates well with the results of organoleptic panels in terms of both raw odour and cooked flavour scores. The instrument is light and portable and provides a non-destructive rapid test which can be applied in the market place.

Frozen fish deteriorate on storage at rates depending on the original quality of the fish, the conditions of storage, and the quality of the packaging. These changes can readily be detected on thawing and cooking, and range from 'freezer-burn' (surface dehydration due to inadequate packaging), 'drip' (loss of liquor from the tissue on thawing), dryness and toughness on cooking and the development of a 'cold store' flavour. Oxidative rancidity is an additional problem with fatty fish. A range of both chemical and physical methods has been proposed to follow and monitor these deteriorative changes. None has so far proved wholly satisfactory, while some have proved useful for some species of fish and certain practical situations. Sensory methods still provide the fundamental point of reference and are likely to continue to do so in the immediate future. [see Mills, A (1975) *J. Food Technol. 10*, 483–496.]

4.10. Fish farming

Simple forms of fish farming have been practised for centuries. In Europe 'stew' ponds were used in pre-Reformation times to provide fish for the monasteries' Friday meals when the eating of meat was forbidden. In the Far East, flooded paddy-fields have been used to cultivate the grass-carp and similar species. To western eyes it is still strange to see a group of young men fishing with rod and line in a field green with rice and, what is more, catching fish! However, such methods were comparatively primitive, the yields per unit area of water surface were low and the fish crop uncertain.

Since the commercially prized species of fish have become scarcer and more highly priced, there has been renewed scientific and commercial interest in fish farming especially in the United Kingdom, Israel, Denmark, the United States and Japan. The world production of farmed fish and shellfish has now risen to around 6,000,000 tonnes per annum.

To farm the seas directly, as the stock farmer does the land, it is necesssary to do as he does. Selected areas such as off-shore waters or estuaries would be enclosed, the stock would be moved from breeding to fattening areas, weeds, predators and diseases would have to be controlled, fertilizers would be used to boost supplies of the right foods and, finally, selective breeding would be used to improve eating quality and the efficiency of feed conversion.

Enclosing an area of sea-water, even where local geographical features (such as the sea-lochs of Scotland or the fjords of Norway) appear to offer natural advantages, presents formidable practical difficulties, as Scottish experience has shown. Although ways and means of doing this will come in time, the present main lines of development aim to exploit waste heat from power stations to speed growth rates in tanks of sea-water, when rates of up to double those observed in wild fish have been obtained.

The cultivation of fish of the *salmo* species (salmon, sea-trout, rainbow trout and brown trout) has been undertaken over many years for the primary purpose of restocking rivers and lakes. The methods devised and experience gained has more recently been adapted to producing fish of commercial size for the retail market. In fact, fish farming in fresh water is making more rapid progress than marine farming although, in the long term, marine farming offers much greater commercial rewards when the technical problems have been overcome.

Further reference

The Food Resources of the Ocean, Holt, S. J., in *Readings from Scientific American, Food*' (1973) pp. 139–151, Freeman, San Francisco.

The Biology of Sea Fisheries, Wimpenny, R. S., in *The Nation's Food* (1946) Bacharach, A. L. and Rendle, T. (eds), The Society of Chemical Industry, London.

Animal Foods, in *Recent Advances in Food Science* (1962), Hawthorn, J. and Leitch, J. M. (eds), vol. 1, pp. 129–201, Butterworth, London.

Fish as Food (1961–1965) Borgstrom, G. (ed.), vol. 1–4, Academic Press, London and New York.

Fish and Fish Products, Cutting, C. L. and Spencer, R., in *Quality Control in the*

Food Industry (1968) Herschdoerfer, S. M. (ed.), vol. 2, pp. 303–353, Academic Press, London.

Fish Farming, Brown, E. E., in *Encyclopaedia of Food Science* (1978) Peterson, M. S. and Johnson, A. H. (eds), pp. 258–263, Avi Publishing, Westport.

Developments in Handling and Processing Fish (1965) Burgess, G. H. O., Fishing News (Books) Ltd, London.

The Bacteriology of Fresh and Spoiling Fish and the Biochemical Changes Induced by Bacterial Action (1977) Shewman, J. M., Torry Research Station Memoir, no. 546, Torry Research Station, Aberdeen, Scotland.

5

Poultry and Eggs

5.1. Poultry

The domestication of the hen goes back at least three thousand years in the records of ancient Egypt.

All varieties of the domesticated fowl, *Gallus domesticus*, are believed to owe their descent to *Gallus bankiva*, the modern form of the Indian jungle-fowl. In contrast to the ancient history of the fowl, the turkey, *Meleagris gallopavo*, was found in Mexico and the United States by the early settlers as a wild bird, and its domestication is thus relatively recent. It seems to have first reached Europe about 1524. There is some evidence that the Indians of New Mexico had domesticated the bird prior to European settlements.

The contraction mechanisms of mammalian and avian muscle are essentially similar. Thus, after death, lactic acid is produced in chicken muscle which thereafter passes through rigor mortis in essentially the same fashion as does beef. However, the prized tender breast muscle of chicken and turkey is little used by the flightless domesticated forms of these birds and is thus pale in colour and low in myoglobin.

In contrast, the breast meat of wild birds (such as wild duck, geese and grouse) is dark coloured and well supplied with respiratory pigment. As would be expected, the amino acid composition of poultry meat (on a mg/g nitrogen basis) is almost identical to that of beef, both in turn being similar to but not identical with that of fish.

The raw breast meat of domestic chicken and turkey is low in fat, (about 0.7 per cent), the thigh, back, ribs and neck being relatively higher (3–6 per cent). In contrast to beef, lamb and pork, avian meat has much higher levels of polyunsaturated fatty acids, broiler chicken fat having around 14 per cent, turkey about 21 per cent and Scotch grouse (that most delicious of game birds) over 60 per cent polyunsaturates expressed as a percentage of the total fatty acids present.

The vitamin levels in poultry muscle are also similar to those of meat, vitamin C being absent, vitamins A and D present only in traces, and thiamin and riboflavin levels much like those of beef and lamb, while nicotinic acid and B_6 levels are higher in poultry meat. However, for most practical dietary purposes, chicken and animal muscle may be regarded as being of similar nutritional value.

From a technological point of view the small size of a poultry carcase allows more rapid slaughter, evisceration, cooling and preparation for marketing, than do other forms of meat. Modern intensive methods of rearing combined with frozen distribution have so reduced the unit cost of poultry in the shops that consumption rose from 6.5 to 19.6 kg per head per annum in the U.S.A. between 1926 and 1976. A similar situation holds in some other countries. For example, sales of turkeys quadrupled in the United Kingdom between 1960 and 1977. Thus, while in most households poultry was regarded as a luxury forty years ago, it has now become one of the least expensive forms of animal protein.

Modern intensive rearing methods are aimed at the most efficient conversion of poultry food to poultry meat or to eggs.

Thus a bird weighing 1.6 kg live weight will yield about 800 g meat averaging around 20 per cent protein and 7 per cent fat (i.e. 160 g protein and 56 g fat). During its lifetime the bird will consume 3.5 kg of feeds containing 700 g protein. The yield of chicken meat is thus about 23 per cent in terms of protein and about 11 per cent in terms of energy. However, much of the feed used is either unattractive to humans or inedible by them and the hen can therefore be regarded as a convenient device for converting low-grade to high-grade human food.

5.2. Eggs

It is convenient to continue the line of thought of the previous section by considering the efficiency of the hen as a producer of eggs. A modern bird will lay about 250 eggs per year weighing some 15 kg and containing 1.75 kg dry weight of egg protein. To do this the hen consumes 40 kg of feed containing 6.5 kg of protein. Thus it produces 0.38 kg eggs/kg feed or 0.27 g egg protein/g feed protein. When considering these figures it should be remembered that egg protein is digested and assimilated more efficiently than any other protein by man and indeed is used as the reference protein by nutritionists against which all other proteins are compared for biological value.

An average egg weighs about 58 g, has a specific gravity of 1.09 and a surface area of about 68 cm^2. An outer gelatinous cuticle, 10 μm in thickneess, covers the shell, which weighs about 5 g and consists of 96 per cent calcium carbonate, 2 per cent magnesium carbonate and 2 per cent protein. The shell is penetrated by between 7,000 and 17,000 funnel-shaped pores, 6–23 μm at their inner ends and 15–65 μm at their mouths. The outer cuticle obscures the mouths of many but not all of the pores and thus provides a first line of defence against bacterial invasion of the contents. The normal shell thickness is from 300 to 340 μm and on its inner surface is a double fibrous or keratin membrane about 70 μm thick, which separates (usually at the blunt end) to produce an air sac. This is normally quite small when the egg is newly laid but may increase in size on storage due to evaporation of water through the shell pores, which serve as a medium of gaseous exchange for the growing embryo.

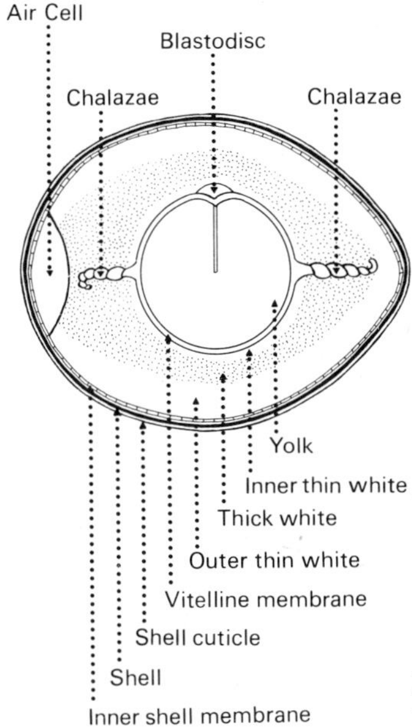

Fig. 5.1. The structure of the avian egg

Shells vary in colour from white to brown, the pigmentation being due to the presence of porphyrins, and its degree is linked to breed, that is to say, to the genetic inheritance of the laying bird. Contrary to popular belief, the colour of the egg bears no relationships to eating quality or nutritional value.

The structure of the avian egg is shown in Fig. 5.1.

Egg-white is visibly non-uniform. The outer or thin white is a moderately viscous liquid surrounding the inner or thick white which has gel-like properties. The thin vitelline membrane surrounds the yolk and this is separated from the thick white by an inner layer of thin white. The yolk is supported by fine strands of fibrous material called the chalazae, which extend through the inner thin white to the thick white and help to support the yolk in the centre of the egg. Integral with the vitelline membrane is the germ cell or blastodisc, from which, if fertilized, the chick later develops.

5.3. The composition of the hen's egg

The chemical composition of the principal components of the hen's egg are shown in Table 5.1.

The protein composition of egg-white is shown in Table 5.2. Egg-white has properties of importance both to the chef and in the baking industry.

Table 5.1. The composition of the egg (edible portion)

Constituent	Whole egg	White	Yolk
	%		
Water	74.5	88.5	47.5
Protein	12.5	10.5	17.4
Lipid	11.8	—	33.0
Free carbohydrate	0.4	0.5	0.2
Ash	0.8	0.5	1.1

Its functional significance is that the globulins (including lysozyme) will produce a foam when whipped in air, which is stabilised by the ovomucin. During cooking ovalbumin is denatured and thus imparts structural rigidity to products such as cakes, meringues and omelettes.

Ovomucin is believed to form a complex with lysozyme which may be responsible for the jelly-like condition of the thick white. The flavoproteins are complexes with riboflavin. Lysozyme, as well as contributing to the stability of the thick white, has antibacterial properties, producing lysis of the cell walls of certain bacteria (usually but not invariably gram-positive), presumably through its enzymic activity in catalysing the hydrolysis of β1–4 glycosidic bonds. Ovomucoid is a trypsin inhibitor, conalbumin can chelate iron, zinc and copper while avidin combines with biotin making this vitamin nutritionally unavailable (see page 32). Fortunately the trypsin-inhibitory properties of ovomucin and the biotin-binding properties of avidin are destroyed by heat denaturation on cooking.

The pH of egg-white is about 7.6 when the egg is laid but this rises to 9.0 or even 9.4 after a few days due to loss of CO_2 through the cell pores. The pH of the yolk is 6.0–6.3 as laid and rises only slowly to an eventual

Table 5.2. Protein composition of egg white

Protein	Percentage of egg-white solids
Ovalbumin	54
Conalbumin	15
Ovomucoid	11
Lysozyme	3.5
Ovomucin	1.5
Unidentified globulins	8
Flavoproteins	0.8
Ovoinhibitor	0.1
Avidin	0.05

Table 5.3. Composition of egg-yolk

Constituent	Percentage of yolk solids
Proteins (about 33% of yolk solids)	
Livetins	4–10
Phosvitin	5–6
Vitellin	4–15
Vitellenin	8–9
Lipids (about 63% of yolk solids)	
Triglycerides	41
Phospholipids	18.5
Cholesterol	3.5
Free glucose	0.2
Inorganic material	2.1
Others	1.5

6.5–6.8. Very few foods are alkaline in reaction and egg-white is therefore unusual in this respect.

The yolk is a complex of lipid and proteins (Table 5.3) which are chemically and functionally related. As used in food yolk behaves as an emulsifying agent, a property which depends on its high phospholipid content and the fact that all of the lipid, including the triglycerides, appears to be associated with at least two of the proteins, vitellin and vitellenin.

The lipid composition of the hen's diet has no influence on the amount of saturated fatty acids (mostly palmitic and stearic and amounting to around 33 per cent of the total fatty acids). However, the feeding of high levels of polyunsaturated fatty acids substantially increases the level of the linoleic acid with a corresponding decrease in oleic acid, which normally accounts for up to 45 per cent of the total fatty acids. While it is convenient (chemically and in nutritional terms) to consider yolk lipids and proteins separately, it must always be remembered that in the intact yolk they are complexed in the form, for example, of lipovitellins and lipovitellenins.

The high cholesterol content has led to a view that severe restriction of egg intake should be practised by any sections of the community believed to be particularly vulnerable to coronary heart disease. However, there is no unanimity of view on the matter.

5.4. Vitamins in eggs

Vitamins A, D, E and K are present in variable amounts, which appear to reflect the levels in the diet of the fowl. The colour of egg-yolks, for example, reflects the carotenoid intake in the feed and may range from almost white to deep orange-red. Normally, the principal compounds found are zeaxanthin and lutein, but if vitamin A precursors, such as β-carotene, are present in the feed some will be transferred to the yolk. Since rich yellow yolks are desired by the public, some careful studies have been made of the transfer of carotenoids from feed to egg. It appears that they are not all transferred with equal efficiency, and knowledge of this is used in compounding feeds to give an attractive yolk colour. Two synthetic compounds, the orange β-apo-8′-carotenoic acid ethyl ester and β-apo-8′-carotenal are used because of the efficiency with which they are transferred and because in admixture with synthetic red canthaxanthin, they produce a desirable shade of yolk colour.

However, as Professor R. A. Morton has pointed out, this is a particularly elegant application of chemistry. If a proportion of either of the first two is metabolised by the hen, vitamin A is a major product. Likewise, if the human consumer metabolises the pigment from the yolks, vitamin A will result. Thus an attractive colour is combined with enhanced nutritional value. On normal diets the vitamin A levels range from 70 to 300 μg retinol/yolk, or 350 to 1850 μg/100 g yolk. The recommended intakes of this vitamin for adolescents and adults are 750 μg/day, so one breakfast egg may provide about one-third of this amount.

Vitamin D in feed appears to be efficiently transferred to yolk and it is normally present at levels around 2.5 μg/100 g yolk. Recommended dietary intakes of 10 μg per day have been recommended for this vitamin, but less is necessary when the subject has adequate access to sunlight.

Vitamin E is normally present at levels of about 75 mg/100 g yolk and vitamin K is also present in moderate quantities. Readers will remember from the first part of this book that primary deficiency of these vitamins is unknown in man. Table 5.4 gives the levels of B-group vitamins in egg.

Table 5.4. B-group vitamins in egg
Mean values are given in mg/100 g components as laid. From Holman, W. I. M. (1956) *Nutr. Abstr. Rev.* *26*, 277–304.

Vitamin	White	Yolk
	mg/100 g	
Thiamin (B_1)	0.002	0.260
Riboflavin (B_2)	0.372	0.442
Nicotinic acid	0.093	0.016
Total B_6 (as pyridoxine)	0.216	0.308
Pantothenic acid	0.304	3.06
Biotin	0.007	0.052
Folic acid	0.0016	0.0232

Vitamin B_{12} is also present in egg-yolk at a level of around 1.8 $\mu g/g$. Eggs contain only a trace of vitamin C.

5.5. Mineral in eggs

The major inorganic constituents in egg are as shown in Table 5.5.

Table 5.5. Mineral composition of eggs

Mineral	mg/100 g
Calcium	60
Chlorine	130
Iron	3
Magnesium	12
Phosphorus	220
Potassium	140
Sodium	130

In summary we can count eggs as contributing protein of high quality to human diet while being a good source of fat-soluble vitamins, and a fair source of water-soluble vitamins and of minerals.

5.6. The storage of eggs

Eggs deteriorate by bacterial and fungal rotting. This may be delayed by storage at a low temperature or by treatment of the shell to seal the pores (for example, with sodium silicate, calcium hydroxide paste, by dipping in light mineral oil or in other propriety products aimed at producing the same effect).

In the absence of microbiological action, deteriorative chemical changes still take place resulting in thinning of the thick white and weakening of the vitelline membrane. A cold-stored egg, when broken out to fry, may appear watery, and lie flat on the plate compared to a fresh egg (see Fig. 5.1). Furthermore, during storage, evaporation of water through the shell takes place resulting in an unsightly enlargement of the air sac.

Evaporation may be controlled by humidifying the storage room to, say, 85 per cent relative humidity, but this gives rise to an increased occurrence of rotting, especially from moulds. This can be controlled by the introduction of carbon dioxide into the atmosphere of the storage room. From this finding, two general methods have evolved. Using a high concentration of carbon dioxide (60 per cent or more) prevents mould growth, and the relative humidity of the chamber can be increased to 96 per cent, thus reducing evaporation to small proportions. These advantages are gained at the expense of rapid thinning of the thick white. In contrast if 2.5 per cent carbon dioxide is used, it is still necessary to control humidity (say at 80 per cent) to prevent mould growth and the rate of evaporation is thus relatively rapid. In compensation, the rate of thinning of the thick white is greatly reduced and the weakening of the vitelline membrane proceeds more slowly than in normal storage.

5.7. The functional properties of egg-white and egg-yolk

Egg-whites are valued for their foaming properties while yolks are particularly useful as emulsifying agents. Egg-whites are therefore characteristically regarded as basic ingredients in the preparation of products such as angel cakes, meringues and dessert souffles, while yolks are used in mayonnaise, Hollandaise sauce, salad dressings and many types of cakes as well in chou (cream-puff) pastry. To some extent both sets of properties are combined in products such as omelettes and sponge cakes. Although similar properties may be observed in other natural products (for example blood and cereal albumins exhibit foaming properties and vegetable oil lecithins are good emulsifying agents), eggs remain the preferred agents for many applications in terms of giving products of high structural quality combined with an attractive flavour.

The foaming properties of egg-white and the volume and stiffness of foam produced are, as would be expected, dependent on the duration and temperature of agitation as well as on the type of beater used. The foaming quality of a given sample is also dependent on (a) the water content of the albumin, (b) the pH, (c) the presence of other substances such as sugar and salt and (d) the absence of contamination from yolk proteins.

Alkalis do not greatly influence foaming but adjustment of the pH to around 6.0 gives optimum results. However, this effect is not a simple one, for the use of potassium hydrogen tartarate as an acidifying agent gives better results than either acetic or tartaric acid. Also the water content of the 'albumin' is important when egg-whites are being whipped and the addition of water up to 33 per cent increases foam volume. Sugar and salt tend to decrease foam stability and increase whipping time. While there have been many studies on the mechanism of foaming and the stability of foams once formed, and while there is general agreement that the globulin fraction of egg-white is required for the production of the foam, and ovomucin and ovalbumin for its stability and structure respectively, a definitive explanation of foaming behaviour is not yet available.

An important problem in industrial egg processing is separating whites and yolks, since yolk proteins and yolk lipids have a marked effect in reducing foam volume. During prolonged storage of egg-yolk lipids migrate into the albumin and the functional properties of the albumin may therefore deteriorate.

Further reference

Poultry Production, Wilson, W. O., in *Readings from Scientific American, Food* (1973) pp. 131–138, Freeman, San Francisco.

Egg Quality (1968) Carter, T. C. (ed.), Oliver and Boyd, Edinburgh.

The Chemical Composition of Eggs, Parkinson, T. L. (1966) *J. Sci. Food Agric. 17*, 101–111.

Eggs as a Source of Protein, Vadehra, D. V. and Rath, K. R. (1973) *Crit. Rev. Food Technol. 4*(2) 193–309.

6

Milk and Milk Products

6.1. Introduction

Used for its original purpose of nourishing newly born animals, milk is a near-perfect food designed for transfer from producer to user in the most convenient and hygienic fashion imaginable. When human intervention diverts it from its original purpose of feeding calves, lambs, kids, foals and the progeny of the buffalo and camel to feeding mankind and his offspring, it is not unreasonable to expect complications. That is what this chapter is about.

While historically man has used the milk of almost every domesticated animal for human food, cow's milk now far outstrips all the others in importance. Modern dairying is a recent development. Until the growth of large cities made it impossible, many (perhaps even most) families in the United Kingdom and in other temperate climates kept a cow. Even in London and other large cities it was easier to keep cows in the city and carry fodder to them, than to transport milk from the countryside. This situation only changed in the second half of the nineteenth century when the development of railway systems made possible the rapid transport of milk from farm to city.

Used by a suckling infant, whether animal or human, milk is a very safe food. However, it is also an exceptionally fine medium for bacterial growth alike for pathogens and spoilage organisms. Milk technology has its roots in the combating of this effect.

6.2. The chemical composition of milk (Table 6.1)

The chemical composition of milk is influenced by the following factors.

1. The animal producing it: human milk, for example, is low in protein (1.6 per cent) and high in lactose (7.0 per cent). A newborn infant doubles its weight in 180 days. Pig's milk is high in protein (7.2 per cent) and low in lactose (3.3 per cent) and the newborn piglet doubles its weight in 14 days. Cow's milk is intermediate in that it contains around 3.4 per cent protein and 4.8 per cent lactose and the newborn calf requires about 47 days to double its birth weight.

2. Inherited variations: the different breeds of dairy cattle have characteristic differences in the composition of the milk they produce, Holsteins and Ayrshires tending to produce milk lower in fat and protein compared with, for example, Jersey's and Guernseys. Differences between cattle varieties are not nearly as great as from genus to genus.

Table 6.1. Typical composition of pasture-fed English milk
Energy value 66 kcal/100 g; 0.28 MJ/100 g

Constituent	
	g/100 g
Water	87.6
Total solids, comprising	12.4
Protein	3.3
Fat	3.6
Lactose	4.7
Calcium	0.12
Phosphorus	0.10
Magnesium	0.01
Iron	0.03×10^{-3}
	μg/100 g
Vitamins	
A (as retinol)	50
D (as cholecalciferol)	0.1–0.4[a]
C	2000
B_1	45
Riboflavin	150
Nicotinic acid	80–400
Pantotheric acid	400
B_6 (pyridoxine)	30
Biotin	5
Inositol	14
B_{12} (cyanocobalamin)	7

[a] The levels of this vitamin in milk vary from winter (low) to summer (high) and the figures quoted represent this range.

3. The feed of the cow has comparatively little effect on the major constituents of the milk, but minor components, such as vitamin A, iodine and some trace metals, are decidely affected by it.

4. Seasonal variations: the fat content of milk tends to be higher in autumn and winter than in spring and summer and the protein content follows the same trend.

5. Age: the fat content of milk tends to fall as a given cow becomes older.

6. Stage of lactation: the first milk produced after a calf is born is quite different from later milk, being richer in minerals and protein and poorer in lactose. It is known by the term ‘colostrum’. Fat, protein and solids-not-fat tend to decline for six or seven weeks and then rise again towards the end of the lactation.

7. Disease: infections of the udder tend to result in a lowering of milk fat content, lactose and casein and an increase in serum proteins.

8. Climatic conditions: when the temperature of the cow's environment rises above 21°C, she tends to eat less and to produce less milk. When a shade temperature of 30°C is reached, the food ingested drops to a half or even a third of that consumed at 15°C, and the milk yield may be reduced by 70 per cent or more. This is clearly of major importance in the tropics and semi-tropics and attempts are being made to improve milk yields under these conditions by cross-breeding European and North American strains with local tropical cattle.

One pint of milk supplies seven-eights of the calcium, one-quarter of the protein, about one-third of the riboflavin and one-fifth of the vitamin A required by a moderately active man. For a child the contributions are not only larger but more uniform, illustrating the special value of milk for the young. Unfortunately the vitamin C in milk is unstable and is readily destroyed by heat, by the catalytic action of traces of copper in the milk and by exposure to bright sunlight on the doorstep after delivery. Milk is not regarded as a reliable or important source of this vitamin in human diet.

6.3. The physical properties of milk

The ivory colour of milk may become blue-white on dilution with water or skimming. The naturally yellowish colour of the fat is associated with the presence of carotenoid pigments in the feed of the cattle. Although milk as produced from the cow has a delicate rather than a strong flavour, milk fat readily absorbs foreign odours and thus care must be taken on its storage.

The specific gravity of milk is normally between 1.027 and 1.035. Since the density of milk 'serum' is about 1.035 g/cm^3 it is clear that increasing milk fat will tend to reduce specific gravity.

The minimum solids-not-fat figure is usually accepted as being about 8.5 per cent and if the figure falls below this the question of whether the milk has been watered either as a deliberate fraud or by accidential spillage may arise. Nevertheless, genuine milk samples are found from time to time which fall below this figure. Since the osmotic pressure of milk is linked to the osmotic pressure of the blood from which it receives its constituents, osmotic pressure provides a more constant factor than solids-not-fat values. It is conveniently measured by using the freezing point of milk as its indicator and the delicate Horvet technique, which can detect freezing-point differences of less than one-thousandth of a degree, is widely used for this purpose. Virtually all genuine milk samples fall within the range −0.530°C and −0.550°C, and a freezing-point higher than the first of these two figures provides strong evidence of the addition of water. Naturally milk does not contain nitrate but many water supplies do, so the presence of a trace of nitrate in milk provides confirmatory evidence.

The size of the fat globules in fresh milk which has not been subject to agitation varies from 0.1 μm to 10 μm, the average being about 3 μm. The breed of the animal and the stage of lactation are the main factors affecting globule size, Jersey cows, for example, having larger fat globules in their milk than Ayrshires. The larger the fat globules the greater the ease of churning. The globules are stabilised by a phospholipid-protein layer or layers commonly referred to as the fat-globule membrane. While there is some degree of agreement about the nature of the materials present in the membrane, their organisation remains a matter of debate. The nature of the membrane structure is of practical importance in its influence on the technology of many aspects of milk processing, including cheese making, butter making and operations such as drying whole milk.

6.4. Milk proteins

When acid is added to milk some of its protein precipitates in the form of a curd-like clot, while the remainder remains dissolved in the liquid, known as whey or milk serum, which separates. This process occurs spontaneously when raw milk is allowed to stand for a number of hours in a warm place owing to the action of lactic-acid-producing bacteria growing in the milk. A similar, but not identical reaction is produced by the action of rennet which is extractable from calves stomachs by salt solutions and which contains the enzyme rennin. The action of rennin on milk marks the first stage in its digestion by the young calf. Equally, when applied for human purposes, the action of rennin is a key operation in the production of most cheeses, although a few types of cheese are also made from the acid-precipitated curd of sour milk.

The precipitated fraction, or curd, is known as casein (or whole casein) and consists of a mixture of proteins all of which contain phosphate groups. They can be separated (for example, by electrophoresis) into alpha (α), beta (β) and gamma (γ) fractions with α contributing about 66 per cent of the total. α-Casein is now recognised as a mixture of proteins with differing functional properties. These include the α_s caseins which are coagulated by calcium ions, and κ-casein, which is not calcium sensitive and which stabilises the casein micelle.

Casein micelles are spherical bodies (ranging in size from about 40 nm to 300 nm) suspended in milk serum and which give the characteristic colour and appearance to milk. The structure of the micelle is not fully understood but it probably consists of an inner core of α_s and β calcium caseinates surrounded by an outer shell of κ-casein, which prevents the precipitation of the core proteins by the calcium ions present in the serum. The α_s caseins have been subdivided into four principal groups, at least one of which (α_{s_1}) varies in its detailed structure with the genetics of the cow producing it. Casein chemistry is therefore complex in detail, but its biochemical and economic importance justifies the scientific attention which it has received.

The whey proteins remaining in solution after the casein has been removed include albumins and globulins, and a number of enzymes and minor proteins of unknown function. The major constituent of whey proteins is β-lactoglobulin which occurs in three genetic variants differing in their composition and properties. α-Lactalbumin is next in importance and its function is associated with the enzyme lactose synthetase. Whey proteins also include a proteose/peptone fraction and an antibody fraction consisting of immunoglobulins. These immunoglobulins are present in colostrum at higher levels than in normal milk. They may be assumed to protect the newborn animal against environmental infections, since there is ample evidence that they can be transferred from the gut to blood.

Milk contains a number of enzymes, most of which appear to be present as escapees from the mammal during secretion rather than as functional constituents. For example, several phosphatases are present and an alkaline phosphatase which is always present in raw milk is destroyed by the time/temperature combination required for efficient pasteurisation. Its presence or absence is the basis of the well-known phosphatase test for the effectiveness of a pasteurisation process.

Milk lipases may produce undesirable flavours in stored milk, in butter and in products such as icecream mixes. The bitter taints produced stem from the hydrolysis of milk triglycerides containing short-chain fatty acids. On the other hand the same lipases are believed to contribute to the desirable flavours of certain cheeses. Other enzymes present include peroxidases, which some believe to be responsible for 'oxidised' flavours in milk, and a casein-hydrolysing enzyme which may be of significance in cheese manufacture and which is active under slightly alkaline conditions.

6.5. Milk lipids

The lipids associated with the milk fat globule are mostly triglycerides, the other lipids amounting to only about 1.5 per cent of the total and of this around 0.5 per cent is diglyceride or monoglyceride, the balance being made up of phospholipids, sterols, free fatty acids and traces of other substances such as cerebrosides, squalene (a precursor of cholesterol), waxes and the fat-soluble vitamins.

The fatty acids of milk triglycerides are unusual. For example, over 9 per cent consist of the short-chain fatty acids, butyric, caproic, caprylic and capric (C_4, C_6, C_8 and C_{10} respectively). About 40 per cent consists of the saturated C_{16} palmitic and C_{18} stearic acids, oleic acid provides about 30 per cent of the total, the polyunsaturates linoleic and linolenic add a mere 3 per cent, the remainder being mainly lauric (C_{12}) and myristic (C_{14}) with smaller amounts of unusual odd-numbered or branched fatty acids. The presence of short-chain fatty acids form the basis of the famous Reichert-Meissl test, which measures volatile fatty acids, and is used to distinguish butter from other fats such as margarine.

6.6. Lactose

The characteristic carbohydrate of mammalian milks is lactose and non-milk sources of lactose are rare. It is less well-known that milk may contain traces of other sugars but these do not appear to be of any practical significance to the food scientist.

The mucosal cells of the small intestine secrete the enzyme lactase which converts lactose to its hexose monosaccharide constituents, glucose and galactose. If, for any reason, this enzyme is absent, lactose passes unchanged into the large intestine where it is fermented by the intestinal flora, producing loose stools containing lactic acid. The enzyme is present in healthy infants and continues into adult life in communities which habitually use milk as food. In Asian and many African countries where milk is only fed to infants and very young children, the ability to produce lactase is lost in later life and cannot be regenerated. Enthusiasm for milk diets must thus be conditioned by appreciation of the fact that many races cannot drink milk in volume without the development of unpleasant digestive symptoms.

6.7. Bacteriological quality of milk

As it flows from a healthy udder, milk is sterile but it is immediately contaminated by contact with air, by the exterior surfaces of teat and udder and (however careful the hygiene observed in the milking parlour) by the pipelines and holding tanks of the milking parlour. It is clearly in the users' interests that the milk be free from pathogens and as free from spoilage organisms as is consistent with economic handling and distribution. Storage life is a function of these factors, and this is ensured by the almost universal application of pasteurising or sterilising processes to modern supplies of milk. The procedures used consist of holding the milk at 65–68°C for 30 minutes or at 72°C for not less than 15 seconds. Both treatments destroy about 99 per cent of the initial contamination, hardly or do not affect flavour, and destroy 10–20 per cent of the vitamin B_1 and C contents. However, one important point should be noted here. Raw milk, exposed to the air at room temperature, becomes rapidly infected with lactobacilli which, by producing lactic acid, sour the milk and precipitate the casein. In warm weather the milk may sour in 18 hours if not refrigerated.

Pasteurising destroys the acid-formers and when the milk eventually spoils (perhaps after seven to ten days at refrigerator temperature) the course of the spoilage is different with proteolytic forms predominating and the milk becoming foul-smelling and 'rotten'.

Prior to the near-universal application of pasteurisation, milk was a major cause of infective illness. For example, as recently as the nineteen thirties in Britain there were around four thousand new cases of tuberculosis each year caused by bovine tubercule bacilli; pasteurisation of milk and the eradication of infected cattle has produced such a dramatic drop in the incidence that tuberculosis hospital consultants have become underemployed and hospital beds once reserved for tuberculosis patients have been reallocated to other purposes. The virtual eradication of this crippling and often fatal disease is a fine example of collaboration between health authorities, food scientists and technologists and the farming interests involved.

Other human diseases associated with infected milk are undulant fever (*Brucella abortus*), food poisoning (from udder infections of cows collectively referred to as mastitis, and including such unpleasant groups as staphylococci, haemolytic streptococci and salmonellae), diphtheria and typhoid and paratyphoid fevers. Wide spread outbreaks of paratyphoid fever have been traced to contaminated milk. For example, one outbreak involved the infection of more than one thousand people from drinking raw milk from cows excreting *S. paratyphi B*

6.8. Dairy products

Cream, butter and cheese are traditional dairy products whose history goes back to pre-Roman times. Modern production methods are based on traditional principles but are so mechanised that they are best considered as food technologies. This is even more applicable to products such as roller or spray-dried milk, long-life milk, icecream, yoghurt and the like, which while products in their own right stem directly from milk as the sole or principal source of origin. Their production involves sophisticated techniques whose nature is controlled by the composition and properties of the milks which form their starting point.

Further reference

Milk Production and Control (1967) Harvey, W. C. and Hill, H., 4th edn, Lewis, London.

Milk Proteins (1971) McKenzie, H. A. (ed.), vol. 1 and 2, Academic, London.

A Dictionary of Dairying (1963, with supplement of 1965) Davis, J. G., 2nd edn, Leonard Hill, London. (For much detailed and authoritative information on milk production, milk products and the dairy industry.)

The physiology of Milk Secretion in Relation to Milk Composition, Smith, J. A. B., in *Recent Advances in Food Science* (1962) Hawthorn, J. and Leitch, J. M. (eds), vol. 1, pp. 113–121, Butterworth, London.

Milk and Milk Products, Palumbo, Mary, in *Food Theory and Applications* (1972) Paul, Pauline C. and Palmer, Helen H. (eds), pp. 563–611, John Wiley, New York.

Food Poisoning and Food Hygiene (1974) Hobbs, Betty C., pp. 112–118, Edward Arnold, London.

7

Fruits and Vegetables

7.1. Introduction

The commodities so far considered in this volume make major contributions to human diet which can be conveniently grouped by their similarity of composition and properties. Fruit and vegetables, on the other hand, are so numerous in variety, varied in composition, and widespread in distribution, that is difficult to impose order on their study. Perhaps the first thing to note is that few of them are staple foods. Cabbages or oranges are valued supplements to a mixed diet but the propect of obtaining a quarter of one's calorie requirements from either of these would be unacceptable and impracticable. As a rule, fruits and vegetables provide the extras by way of minerals and vitamins which help to balance dietary intakes and which give pleasure and variety in so doing.

But most rules have exceptions. In some parts of the world potatoes form a proportion of diet sufficiently large as to justify their inclusion in a list of staple foodstuffs. In others cassava (also known as manioc) and yam come into the same category. These we shall therefore consider first. From this baseline the numerous other fruits and vegetables used for food can be seen in perspective. The sugar-producing plants, such as sugar-cane and sugar-beet, might also claim a place as major commodities in this classification. Since they are used almost exclusively in the form of refined sugar, which few nutritionists regard as a true food but merely a source of sweetness and calories, they shall only have minor comment.

Furthermore, we shall distinguish fruits and vegetables in culinary rather than in botanical terms. Our present purpose is served by regarding tomatoes as vegetables and rhubarb as a fruit!

7.2. Potatoes

It is no accident that the potato has attained a preeminent position in temperate zones throughout the world. It is a member of the Solanaceae. Some will find it curious that its close relatives, henbane, deadly nightshade and nicotiana are sources of virulent poisons. Potatoes themselves contain modest amounts of toxic glycosides the principal representative being solanine, a compound structured from the alkaloid solanidine linked to the sugars glucose, galactose and rhamnose. Solanine occurs along with other solanidines of the same basic structure but linked to

different sugars

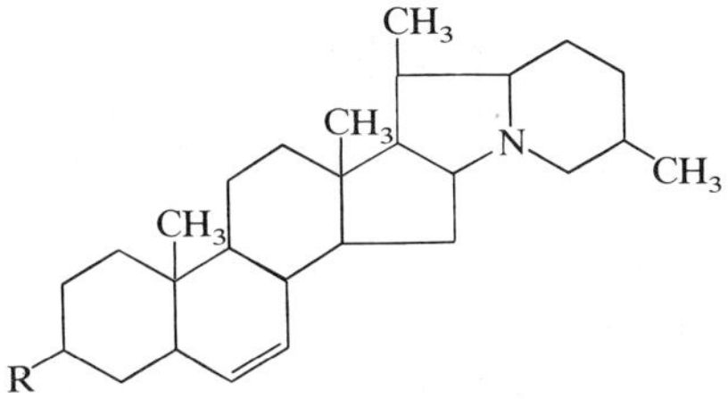

CH₃

R = glucose-galactose-rhamnose-link

Solanine

Normal wholesome potato tubers contain up to about 90 parts per million. Outbreaks of poisoning have been reported from the consumption of potatoes containing from 230 to 500 parts per million. At the same time other cholinesterase inhibitors are present in potatoes and may contribute to the toxic effect. Abnormally high solanine contents have been reported from potatoes exposed to sunlight, and it is advisable, therefore, to avoid eating potatoes which show green patches. Potato sprouts contain high levels, and there is rather more in the skin that in the flesh. Peeling is therefore helpful, and some solanine is also leached on cooking. The presence of traces of toxic substances in foodstuffs is not uncommon and the human body appears capable of coping with small amounts through its detoxication mechanisms. Despite their solanine content, potatoes play a special part in the diet of many countries because of their versatility as a raw material for the cook, their cheapness and their nutrient characteristics.

With the exception of the potato, almost all important vegetables are propagated from seed. As a result of this, crops tend to recede more and more from a selected type and care is required to keep the desired characteristics dominant. The potato tuber permits vegetative propagation, and however hybrid it may be for specific characteristics, the plant reproduces itself without variation from tubers.

This fixation of characteristics combined with high yields and the ease with which potatoes may be stored throughout the winter gives them a unique place in agriculture. Their high food value and cheapness give them a special place in European and North American diet.

7.3. Chemical composition of the potato

Despite the existence of an extensive literature on the chemical composition of the potato, a clear picture is difficult to obtain, partly because of inherent variability of variety, cultural environment, maturity at harvest and conditions of subsequent storage, and partly because of differences in the analytical techniques used by different investigators. The situation is further complicated by the existence of variations in composition of different zones within a single tuber, the solids content increasing from skin to cortex and decreasing from the cortex to the inner medulla or pith. Table 7.1, assembled from a range of sources, is intended therefore to illustrate the general characteristics of the potato as reflected by reported analyses.

Although traces of carotenoid pigments have been reported in the potato, for practical purposes they are best regarded as not making any contribution to intake of vitamin A. The other fat-soluble vitamins are absent.

About half of the nitrogen present is in the form of protein, the remainder being mainly amino acids together with a few peptides.

The starch granules in the potato are larger in size than those of the

Table 7.1. Composition of whole potato tubers including skin

Constituent	
	g/100 g
Water	76[a]
Protein and amino acids	2.2
Fat	0.1
Total carbohydrate	19.0
Reducing sugars	0.6
Sucrose	0.4
Fibre	0.6
Pectins	0.6
Ash	1.0
	mg/100 g
Iron	0.8
Calcium	7.0
Phosphorus	40.0
Magnesium	24.0
Ascorbic acid	25.0
Thiamine	0.11
Riboflavin	0.04
Nicotinic acid	1.2
Pyridoxine	0.13
Solanine	9
pH	5.6–6.2

[a] The water content of potatoes is very variable, values as low as 66 g/100 g and as high as 85 g/100 g having been reported.

cereals, being typically 60–100 μm in diameter. They tend to be smaller under poor growing conditions such as drought, potassium deficiency etc. It is believed that the size and size distribution of the granules contribute to the quality of the cooked potato. The main components are amylose and amylopectin present normally in a ratio of 1 to 3. Small quantities of phosphate are esterified with the hydroxyl of the C-6 position of a glucosyl residue of the amylopectin, about one in every two hundred of such residues being so esterified. The viscosity of gels from potato starch appears to be correlated with their phosphate content. In contrast, grain starches do not contain phosphate esters. Potato starch is an article of commerce, valued for the viscosity of its gels. It is prepared as a manufacturing by-product of processes such as the production of crisps or from undersized or damaged tubers.

The sugar content of potatoes influences their suitability for certain uses, especially those involving frying processes in deep fat where high sugar contents lead to unacceptably brown colours. The quantity of sugars present in potatoes is greatly influenced by conditions of storage when any or all of the following changes may take place:

a) starch is converted into sugars;
b) sugars are converted into starch;
c) sugars are metabolised to carbon dioxide and water.

When storage temperatures are reduced the rates of all these reactions are decreased, but unequally, and sugars accumulate. By holding the stock for a few weeks at higher temperatures around 15–18°C the sugar level drops. However, the undesirable reactions associated with high sugar content in French fries, crisps and dehydrated potatoes are correlated more closely with reducing sugar levels than with total sugars, thus pointing to the Maillard reaction (between the aldehyde groups of the reducing sugars and the free amino groups of the amino acids) rather than sugar caramelisation as the origin of the trouble.

In general, potatoes containing more than 2 per cent reducing sugars on a dry-weight basis are considered unacceptable for processing. Raw materials of low browning tendency are obtained by using varieties which are poor sugar formers, and by conditioning them for two or three weeks at around 18°C before use.

There is evidence that the pectin content of potatoes is linked in a rather complex relationship (involving also the starch) with the texture of the cooked product. The formation of calcium pectate has a toughening effect on potato texture and this is exploited by blanching potatoes in dilute solutions of calcium ions prior to canning to prevent the breakdown which would otherwise occur due to the severity of the sterilising process.

7.4. Nutritive value of the potato

A first glance at the analysis of the potato suggests that it is of rather low nutritional value compared to cereals. Table 7.2 shows that on a dry-solids basis such a conclusion is not justified. The potato is certainly lower in protein than wheat, but it contains substantial amounts of vitamin C which is entirely absent from wheat. Indeed, potatoes mashed with a little butter and milk to balance the protein and provide fat-soluble vitamins make a well-balanced meal which was for long the basic diet of peasant populations of Ireland and elsewhere.

It is also interesting to compare potatoes with wheat on a basis of yield of nutrients per unit area of land. A good yield of maincrop potatoes is 30 tonnes per hectare, and of wheat 6 tonnes, which represents 7.2 and 5.1 tonnes of dry matter respectively, using the figures in Tables 7.1 and 7.2 as a basis for calculation.

Table 7.3 shows the amounts of the principal components produced per hectare assuming the above yields.

Wheat clearly scores on protein yield but potatoes win on vitamins and minerals, the contribution of vitamin C being especially significant, despite the losses on storage which may be as large as 10 per cent per month. Potatoes also score heavily on energy yield per unit area. In situations where diets are marginal or insufficient to meet energy requirements, a crop which yields more energy per unit area has an advantage. This is enhanced in the case of the potato by the ease and simplicity of cooking in contrast to the comparatively complex arrangements needed to mill and bake cereals. The ease of cultivation of the potato, the simplicity of planting and harvesting which enabled a peasant, single-handed and in

Table 7.2. A comparison of the composition of potato and wheat on a dry-solids basis

Constituent	Potato	Whole wheat
	g/100 g	
Protein	9.2	16
Fat	0.4	2.0
Carbohydrate as starch	79.2	74.1
Fibre and pectins	5.0	2.6
	mg/100 g	
Calcium	29.2	32.0
Iron	3.3	4.5
Ascorbic acid	104	0
Thiamin	0.45	0.41
Riboflavin	0.17	0.20
Nicotinic acid	5.0	6.5

Table 7.3. Crop yield: a compositional comparison between potato and wheat

Constituent	Potato	Whole wheat
	kg/ha	
Protein	662	816
Fat	29	102
Carbohydrate	5702	3779
Fibre and pectins	360	133
Calcium	2.1	1.6
Iron	0.24	0.23
Ascorbic acid	7.4	0
Thiamin	0.0324	0.0209
Riboflavin	0.0122	0.0102
Nictotinic acid	0.36	0.33
Energy (calculated) (MJ)	103×10^3	77×10^3

his spare time, to cultivate a sufficient area to provide enough for the needs of his family all combined to encourage the rapid spread of the tuber after its first introduction to the United Kingdom, probably via Spain and a little before the year 1600. But it suffered from one serious disadvantage—potato blight.

7.5. Potato blight

Phytophtora infestans is a fungus attacking the potato foliage and tubers in the field at a very rapid and progressive rate, and infected tubers continue to rot on storage and transport. This fungus has played its part in history. In Ireland the poor rural communities of the nineteenth century had become dependent on the potato crop as their principal and sometimes their only food. Now a characteristic of blight is the speed at which it spreads in warm humid conditions. In 1845 about half the crop was affected and nothing was then known about how to protect it. The following year the adverse weather conditions again appeared and widespread famine ensued as virtually the whole of the crop rotted in the fields. It is estimated that over a million people perished of starvation. The disaster affected the history of Ireland, England and America. The population of Ireland was reduced from eight million to four million. In England a bitter political struggle ensued from differences in approach to the problem of famine relief (the outcome being so unsuccessful as to be a blot on English history), which ultimately led to the repeal of the Corn Laws. It affected America also by the flood of Irish refugees which eventually managed to emigrate there, and to the continued fermentation of anti-British feelings in that country as a result. Blight remains endemic in Europe and losses from it still occur, although it is now well controlled by spraying with copper salts (as in the well-known Bordeaux mixture) and by dithiocarbamates, warnings being issued by the agricultural authorities when menacing weather conditions threaten. It is worth noting that there are varietal differences in resistance to blight, and it is thus possible to breed for resistance.

Potatoes are subject to other fungal infections of which common scab (*Streptomyces scabies*), early blight (*Alternaria solani*), Fusarium rots, Verticillium wilts are of some significance and also to bacterial attack by ring rot (*Corynebacterium sepedonicum*) and brown rot (*Pseudomonas solanacearum*). Virus diseases are also common and result in great losses of yield. Although all of these cause trouble, the use of clean disease-free seed tubers, good conditions of cultivation with preventative spraying as required, together with care in harvesting and storage now eliminate the possibility of a repetition of a tragedy such as that of the Irish famine.

7.6. Cassava

Cassava or manioc (Fig. 7.1) is the tuber of *Manihot utilissima, Manihot esculenta* and related species. Although a native of South America, the plant was introduced into Africa by the Portuguese during the sixteenth century, where its advantages of locust resistance and resistance to rotting when simply left in the ground after reaching maturity led to its rapid acceptance. It later spread over the tropics and is now a crop of major importance wherever rainfall, soil and climate permit. The world production is over one hundred million tons and growing. Cassava is a heavy cropper, capable of producing up to 60 tonnes per hectare per annum as harvested, although the world average yield is only about 10 tonnes. Low yields are probably due to poor agronomic practice (for example yields rise sharply if weeds are properly controlled), poor varietal selection, inadequate use of fertilizers and attacks by insects and disease.

The shrubs grow to 2–3 m in height and are easily propagated from stem cuttings. The plant is drought tolerant and will still give reasonable yields on poor acid soils. Furthermore, it has no fixed harvest date, which thus gives great flexibility to the farming systems using it.

As harvested the fresh root contains about 60 per cent water, typical analytical figures being as shown in Table 7.4. These, of course, are subject to variation depending on variety and growing conditions.

After reaching maturity at about six months the root may simply be left in the ground until required. However, if left too long it tends to become fibrous. When harvested it must be processed right away or properly stored in clamps or storage boxes to avoid the fairly rapid deterioration which otherwise sets in. About 83 percent of cassava starch consists of

Fig. 7.1. A cassava plant showing the tubers at an intermediate state of development

amylopectin the remainder being amylose. On a dry-weight basis the total starch ranges from 67 to 81 per cent depending on maturity, and in addition 4 per cent of cellulose and 1 per cent of hemicelluloses are present. Free sugars are also present at around 3–5 per cent on a dry weight basis and consist mainly of maltose, glucose, fructose and sucrose.

Table 7.4. Composition of cassava tubers

Constituent	
	g/100 g
Moisture	62
Protein	0.7
Fat	0.7
Carbohydrate	35
Ash	0.8
Fibre	1.0
	mg/100 g
Calcium	25
Iron	0.5
Thiamin	0.05
Riboflavin	0.06
Ascorbic acid	22

Cassava tubers contain about 2.5 per cent lipids (dry-weight basis) of which about half is extractable using conventional solvents. Being regarded as minor constituents, they have not so far attracted much analytical attention. From the work reported, triglycerides appear to make up about half the total lipid, the remainder being the usual mixture of phospholipids and mixed galactosyl glycerides, the overall pattern being not dissimilar to the lipids of the potato. Although both contain over 60 per cent unsaturated fatty acids, the level of 18:2 in potato is more than twice that of cassava in which 18:1 predominates.

All cassava varieties contain the bitter substance linamarin, but are often called bitter or sweet depending on its concentration, which in both cases is highest in or just under the peel. The enzyme linamarinase is also present and when the tuber is cut or grated in preparation for cooking the linamarin is hydrolysed to glucose, acetone and free hydrocyanic acid. This last may reach dangerous levels, figures as high as 700 mg/kg having been reported in the peel although the figures in the tuber proper are much lower, with characteristic values ranging from 10 mg/kg to 90 mg/kg. Since cassava is a staple and consumed in quantity, it is interesting but not altogether surprising to find that traditional cassava products such as garri and konkonte flour contain only low levels of cyanide.

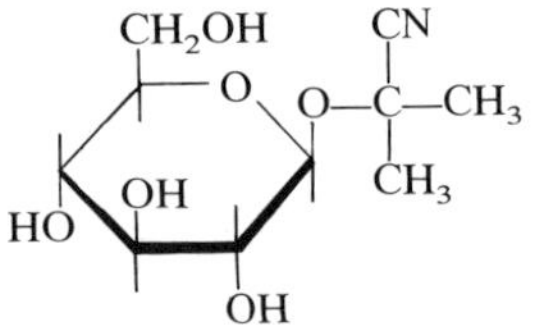

CH_2OH CN

O $O—C—CH_3$

HO   OH CH_3

OH

Linamarin

Since the protein content of cassava is so low, cassava-dependent communities tend to suffer from protein deficiency syndromes, and unless diet is balanced by legumes or some other protein source, such peoples are more than usually vulnerable to disease or toxic responses. For this reason, it seems particularly desirable to ensure low cyanide levels in their staple food.

While there are many local variations in the preparation of cassava products, they all seem to depend on the following stages. Cutting the tubers into strips or grating into shreds releases linamarinase which breaks down linamarin with the release of free hydrocyanic acid, which is subsequently removed by sun drying or fermentation or squeezing out the juice or by an artificial heating process to drive off the HCN. For example, garri is prepared by grating the tubers, fermenting for three days, placing the moist fermented product in sacks, pressing out the juice, then lightly frying or roasting. The resulting product is dried and screened. The product known in the West as tapioca is almost pure cassava starch and is prepared without fermentation.

Cassava has become an item of international trade during the past quarter century as a stock material for blending animal feeds, the E.E.C. countries being particularly active importers, and Thailand an increasingly important exporter. The Thai peoples are rice-eating, and the plant seems to have been introduced there (and also to Indonesia) about fifty years ago to provide a crop on land unsuited for rice production. After harvesting, the tubers are mechanically cut into chips, spread on concrete drying floors in the sun before finally pelleting for export to animal feed compounders. Thailand alone exports over a million tonnes of this commodity each year and it is thus an important article of trade.

7.7. Yams

Yams are the tubers of the climbing plants *Discorea alata, Discorea esculentia* and related species and often grow to about the size of a football. They are grown throughout the wetter parts of the tropics taking about eight months to mature after propagation from cuttings. The tubers contain varying amounts of a poisonous alkaloid, dioscorine, which is easily destroyed by boiling. Fortunately, raw yams have an unpleasant effect on the mouth and throat and are always eaten cooked in one form or another. In some areas they are consumed in sufficient quantity to justify their claim as a vegetable staple.

7.8. The pulses

Although peas, beans, ground-nuts and lentils are widely used in Europe and America they are by no means dietary staples in these areas. In many parts of the tropics, in contrast, pulses play a very important nutritional role which can only be dealt with briefly here.

The Leguminosae form a huge family of about 600 genera with a correspondingly large number of associated species. About forty of these are of economic importance for human food. A few species provide green vegetables in the form of cooked leaves or immature pods, such as the 'mange-tous' dwarf sugar peas, which are growing in popularity in Europe and whose tender pods are lightly cooked at an immature stage. These form a tiny proportion of the total crop, the bulk of which is consumed in the form of dry seeds. Food legumes used as dry seeds are commonly referred to as 'grain legumes' or 'pulses'. The annual production figures are said to be around 60 million metric tons (tonnes) of soya beans, 20 million tonnes of ground-nuts (peanuts) and 10 million tonnes each of peas and beans. However, peas and beans are so widely grown as domestic rather than as agricultural crops that figures quoted may be underestimates. About half of the world crop of pulses is produced in developing countries, where they are of special dietary importance in supplementing cereal diets with essential amino acids and improving nutrition where animal protein is scarce.

The pulses contain 20 per cent or more of protein in a dry-weight basis. The protein is deficient in the sulphur-containing amino acids, but rich in lysine in which cereals are deficient. A combinaton of cereals and pulses may produce a diet of nutritive value approaching that of animal proteins. Furthermore the pulses are relatively good sources of the B-group vitamins except riboflavin. In dry form they are devoid of vitamin C but produce large amounts on germination. Bean sprouts are not only pleasant to eat but are an excellent form of protection against scurvy.

The soya bean (*Glycine hispida*) is the most important member of the family in terms of international trade. They are remarkable for their high protein (40 per cent) and high fat content (20 per cent). The plant is a native of China but is now also grown in the United States, India and Africa. Soya bean oil is a major constituent of margarine while the press cake is an important human food. Soya is the basis of a wide range of Japanese and Chinese dishes, the name soya being derived from the Japanese term 'shoyu' meaning a sauce prepared from the beans. In the west, soya protein is used in the production of meat analogues which may well be of increasing importance in human diet during the next twenty years.

The ground-nut (*Arachis hypogoea*) is also known as the monkey-nut or peanut. The plants are grown all over the tropics. Since the pods develop and ripen underground they have to be harvested by digging. They resemble the other pulses in composition except for their high oil content which may reach 40 per cent. The oil is prized for cooking and margarine manufacture and the extracted cake is an important cattle food.

The consumption of pulses induces flatulence which may be so severe as to discourage their use despite their nutritional advantages. The factors responsible are a matter of some concern to plant breeders who wish to develop varieties in which this disadvantage is reduced or eliminated. The flatulence factors in white beans can be extracted by 60 per cent ethanol in water and the extract contains oligosaccharides including raffinose and stachyose. Both of these substances produce flatulence when consumed in isolation but the degree of flatulence is less than that produced by the original bean in corresponding amount. Other factors yet to be identified are believed to be involved.

The pulses contain a number of antinutritional substances and toxins. Trypsin inhibitors and haemagglutinins present in some species are destroyed by the methods of preparation and cooking generally used. Perhaps more serious is the presence of lathyrogens, present in a range of peas of the genus *Lathyrus*, which cause a form of irreversible nervous degeneration ultimately crippling and paralysing the victim. The disease is insidious, since in small quantities the peas are a nutritious and valuable dietary supplement. However, if they become the major source of food over prolonged periods lathyrism develops. A neurotoxin has been isolated from the seeds of *Lathyrus sativus* and has been identified as β-*N*-oxalylamino-1-alanine.

The consumption of broad beans (*Vicia faba*) in quantity is associated with a form of anaemia known as favism. It is due to an inherited metabolic defect of an insufficiency of the enzyme glucose-6-phosphate dehydrogenase, which is aggravated or made apparent by a food toxin present in these beans.

7.9. Other fruits and vegetables

The staples apart, is there anything interesting to say about onions and oranges and all the other products of vegetable rack and fruit bowl? Tables 7.5 and 7.6 have been compiled to show the general pattern of composition of these products and at first glance look both formidable and dull. But the most interesting thing about them is what they do not contain. Fruits and vegetables are devoid of vitamins B_{12} and D, so a vegetarian who sticks to these lists for sustenance must seek the sun to avoid the D deficiency syndrome as well as a source of cyanocobalamin if he or she is to avoid an untimely death from pernicious anaemia. Pure vitamin A (retinol) is rarely present in fruits and vegetables although most contain carotenoid precursors. The table follows the McCance and Widdowson practice of reporting carotene (which will sometimes be mixed carotenoids) with the proviso that the figures be divided by 6 to express them as vitamin A potencies in retinol equivalents. This factor is probably conservative but the data show clearly enough the value of fruits and vegetables as sources of this vitamin.

A glance down the left-hand columns will also persuade you that despite the seeming solidity of some of them, fruits and vegetables are mostly structured water with small amounts of other things associated with it. The quantities listed are per 100 g, which represents a typical serving for most of them, and only avocados, potatoes and fried mushrooms show up well as energy sources. Truth to tell even at that, most of the energy value of fried mushrooms lies in the fat used for frying. Nor are the protein levels exciting compared to those of meat, milk, fish, eggs and cheese. Vitamins and minerals, however, are quite another matter, the vitamin C column immediately catching the eye. These 'watery' foods

Table 7.5. Major constituents of fruits

Recommended daily dietary intakes for the U.K. are as given by the Department of Health and Social Security, 1969, for a sedentary 65-kg man. The figures quoted are for the flesh of the raw fruits as normally eaten. The water content is rounded to the nearest 1 per cent. Carotene values are given as retinol equivalents ×6.

Fruit	Energy/100 g		Water	Protein	Fat	Carbohydrate	Ca	Fe	Carotene	Thiamin	Riboflavin	Nicotinic acid	Vitamin C
	kcal	kJ	g/100 g				mg/100 g						
Apples, fresh	46	196	84	0.3	tr.	11.9	4	0.3	0.03	0.04	0.02	0.1	3
Apricots, fresh	28	117	87	0.6	tr.	6.7	17	0.4	1.5	0.04	0.05	0.6	7
Avocados	223	922	69	4.2	22.2	1.8	15	1.5	0.1	0.10	0.10	1.0	15
Bananas	79	337	71	1.1	0.3	19.2	7	0.4	0.2	0.04	0.07	0.6	10
Blackberries	29	125	82	1.3	tr.	6.4	63	0.9	0.1	0.03	0.04	0.4	20
Cherries	47	201	82	0.6	tr.	11.9	16	0.4	0.12	0.05	0.07	0.3	5
Currants, black	28	121	77	0.9	tr.	6.6	60	1.3	0.2	0.03	0.06	0.3	200
Currants, red	21	89	83	1.1	tr.	4.4	36	1.2	0.07	0.04	0.06	0.1	40
Damsons	38	162	76	0.5	tr.	9.6	24	0.4	0.22	0.1	0.03	0.3	3
Gooseberries	37	157	84	0.6	tr.	9.2	19	0.6	0.18	0.04	0.03	0.3	40
Grapes	51	217	65	0.5	tr.	13.0	4	0.3	tr.	0.03	0.02	0.2	3
Grapefruit	22	95	91	0.6	tr.	5.3	17	0.3	tr.	0.05	0.02	0.2	40
Greengages	47	202	78	0.8	tr.	11.8	17	0.4	0	0.05	0.03	0.4	3
Lemons	7	31	92	0.3	tr.	1.6	8	0.1	tr.	0.02	0.01	0.1	50
Melons (honeydew)	21	90	94	0.6	tr.	5.0	14	0.2	0.1	0.05	0.03	0.5	25
Oranges	35	150	87	0.8	tr.	8.5	41	0.3	0.05	0.10	0.03	0.2	50
Peaches	37	156	86	0.6	tr.	9.1	5	0.4	0.50	0.02	0.05	1.0	8
Pear	41	175	83	0.3	tr.	10.6	8	0.2	0.01	0.03	0.03	0.2	3
Pineapple	46	194	84	0.5	tr.	11.6	12	0.4	0.06	0.08	0.02	0.2	25
Plums (Victoria)	38	164	84	0.6	tr.	9.6	11	0.4	0.22	0.05	0.03	0.5	3
Raspberries	25	105	83	0.9	tr.	5.6	41	1.2	0.08	0.02	0.03	0.4	25
Rhubarb (raw)	6	26	94	0.6	tr.	1.0	100	0.4	0.06	0.01	0.03	0.3	10
Strawberries	26	109	89	0.6	tr.	6.2	22	0.7	0.03	0.02	0.03	0.4	60
Tangerines	34	143	87	0.9	tr.	8.0	42	0.3	0.1	0.07	0.02	0.1	30
Recommended daily	2,600	10,900	—	65 g	—	—	500 mg	10 mg	4.5 mg	1.0 mg	1.7 mg	18 mg	30 mg

Table 7.6. Major constituents of vegetables

Recommended daily dietary intakes for the U.K. are as given by the Department of Health and Social Security, 1969 as for a sedentary 65-kg man. The figures quoted are for the vegetables as normally eaten in the cooked (boiled unless otherwise stated) or raw state. The water content is rounded to the nearest 1 per cent. Carotene values are given as retinol equivalents $\times 6$

Vegetable	Energy/100 g		Water	Protein	Fat	Carbohydrate	Ca	Fe	Carotene	Thiamin	Riboflavin	Nicotinic acid	Vitamin C
	kcal	kJ	g/100 g				mg/100 g						
Beans, runner	19	83	91	1.9	0.2	2.7	22	0.7	0.4	0.03	0.07	0.5	5
Beans, butter	95	405	71	7.1	0.3	17.1	19	1.7	tr.	—	—	—	0
Beetroot	44	189	83	1.8	tr.	9.9	30	0.4	tr.	0.02	0.04	0.1	5
Broccoli tips	18	78	90	3.1	tr.	1.6	76	1.0	2.5	0.06	0.20	0.6	34
Brussels sprouts	18	75	92	2.8	tr.	1.7	25	0.5	0.4	0.06	0.10	0.4	40
Cabbage (winter)	15	66	93	1.7	tr.	2.3	38	0.4	0.3	0.03	0.03	0.2	20
Carrots	19	79	92	0.6	tr.	4.3	37	0.4	12.0	0.05	0.04	0.4	4
Cauliflower	4	40	95	1.6	tr.	0.8	18	0.4	tr.	0.06	0.06	0.4	20
Leeks	24	104	91	1.8	tr.	4.6	61	2.0	tr.	0.07	0.03	0.4	15
Lettuce	12	51	96	1.0	0.4	1.2	23	0.9	1.0	0.07	0.08	0.3	15
Mushroom (fried)	210	863	64	2.2	22.3	0	4	1.3	0	0.07	0.35	3.5	1
Onions (boilded)	13	53	97	0.6	tr.	2.7	24	0.3	0	0.02	0.04	0.1	6
Peas (fresh)	52	223	80	5.0	0.4	7.7	13	1.2	0.3	0.25	0.11	1.5	15
Green peppers (raw)	15	65	94	0.9	0.4	2.2	9	0.4	0.2	tr.	0.03	0.7	100
Potatoes	80	343	81	1.4	0.1	19.7	4	0.3	tr.	0.08	0.03	0.8	10
Spinach	30	128	85	5.1	0.5	1.4	600	4.0	6.0	0.07	0.15	0.4	25
Tomatoes (raw)	14	60	93	0.9	tr.	2.8	13	0.4	0.6	0.06	0.04	0.7	20
Turnip	14	60	95	0.7	0.3	2.3	55	0.4	0	0.03	0.04	0.4	17
Recommended daily dietary intake	2,600	10,900	—	65 g	—	—	500 mg	10 mg	4.5 mg	1.0 mg	1.7 mg	18 mg	30 mg

are our principal source of ascorbic acid and their contribution of other vitamins as well as calcium and iron is by no means negligible. To prevent the tables becoming overlarge folic acid, pantothenic acid and biotin have not been listed, but those are also present in most vegetables and some fruits, although the folic acid levels in fruits are generally rather low. It is worth reminding the reader that while folic acid deficiency is a common problem amongst poor peoples in the tropics, it is rare in western societies except during pregnancy, so pregnant mums should ensure that they get their fair share of brussels sprouts, spinach and broccoli, which are all good sources of this vitamin, and thus avoid one of the causes of anaemia in this otherwise happy state. (Good as these are, liver is an even better source.)

So much then for fruits and vegetables from the nutritional viewpoint of protective foods. To the food technologist the variety and flavour they add to diet is equally important. From the nutritionists viewpoint, onions, leeks, garlic, shallots and other members of the same family have little to offer, but all around the world they are esteemed for their contributions to the flavour of countless dishes. Grapes, which we solemnly give to hospital patients as valued invalid foods are, also a nutritional viewpoint, little more than weak solutions of sugar. Yet grapes produce the almost unlimited range of flavours and bouquets of fine wines. The reader can scan the table and make his own list of the culinary virtues of fruits and vegetables.

The chemistry of these products is by no means as simple as Tables 7.5 and 7.6 may suggest. They contain a range of substances which affect their technical properties and whose importance becomes obvious when they have to be harvested, transported, processed, preserved and distributed. Changes in them may affect appearance and colour, as well as flavour and texture by the time the products reach the intended consumer.

7.10. Colour in fruits and vegetables

The wide range of colours of fruits and vegetables is almost entirely attributable to three classes of plant pigments, the fat-soluble chlorophylls and carotenoids and the water-soluble flavonoids. The details of their chemical structure and biochemical relationships are more appropriate to a biochemistry text than to one on food science and it is assumed that the student is already familiar with these classes of substances. Some of their properties are of special interest to the food scientists and these must therefore be briefly discussed here.

The molecular integrity of the chlorophylls depends on the stability of the magnesium atom coordinated at the centre of the porphyrin ring.

Now most fruits are strongly acid (pH 3.0–3.5) and even green vegetables tend to be slightly acid (pH usually around 6.0–6.8). The magnesium atom is unstable under the influence of heat and is more unstable under acid rather than alkaline conditions. On cooking, therefore, chlorophyll-containing tissues soon lose their attractive bright green colour and change to the dirty olive brown of the decomposition product pheophytin. The rate of change can be retarded by making the cooking water slightly alkaline (by adding, for example, a pinch of baking soda to peas or brussels sprouts and sometimes by rather more sophisticated treatments in industry), but although this results in reducing chlorophyll decomposition it speeds up the decomposition of thiamin and ascorbic acid. For this reason the use of such alkali treatments is not advised, and in practice the best results are obtained by short-cooking times and the use of as little water as possible.

The carotenoid pigments are, of course, structurally related to vitamin A and some are vitamin A precursors. They are responsible for hues as varied as the red colours of tomatoes, rose hips and water melon, the range of orange and yellow pigments of peaches, citrus fruits, peppers, carrots, apricots, maize and the attractive hues of some spices such as saffron. They also occur in green tissue and may be conveyed therefrom by grazing herbivores to animal products such as beef fat and egg-yolk. When extracted from the plants by fat solvents they are generally found to be present in mixtures. The colours of the pure compounds range from the brilliant reds of lycopene (which dominates in tomato and water melon) and capsanthin, to the orange of β-carotene and bixin, and to the gentler yellows of the xanthophylls. The gradations in shades arising from these mixtures explain much of the subtle colouring of the fruits which contain them.

The carotenoids are generally stable to heat processing and cooking but because of their highly unsaturated structures they are particularly susceptible to oxidation. In presence of oxidising fats they are themselves oxidised in a series of coupled reactions which breaks them into smaller colourless fragments. Although this reaction can occur during autoxidation, its results can be rapid when the lipid is attacked by lipoxygenase enzyme systems. Oxidation of carotenoids can also give rise to off-flavours in dehydrated vegetables.

The flavonoid pigments and their related anthocyanin compounds are of widespread distribution in plants. In addition to providing the colours of soft-fruits such as raspberries, blackberries, cherries, red grapes and many others, they are responsible for the almost limitless range of hue of the flowers of decorative plants. Flavonoids are derived from the parent ring structure shown, the ring on the left being designated A and that on the right B.

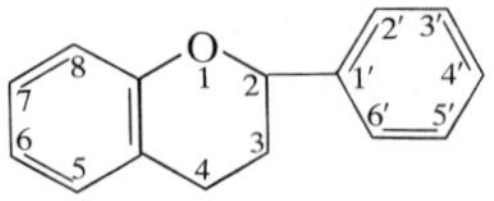

The group takes its name from the flavones which are benzopyrones with the keto group in position 4. When this structure is combined with a hydroxyl on position 3 the resulting group of substances are designated flavonols. When the introduction of a hydrogen atom at 2 gives rise to a saturated ring, the structure is known as a flavanone.

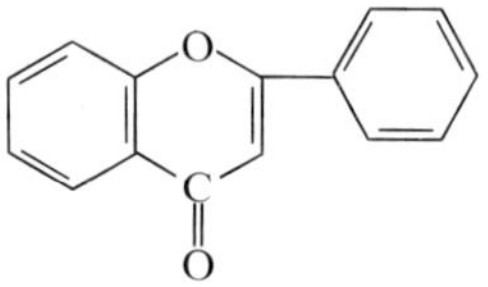

Flavone

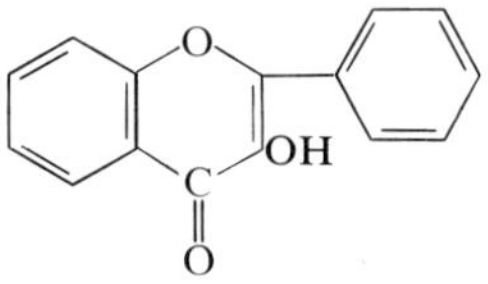

Flavanol

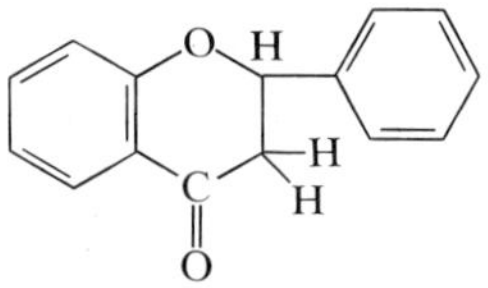

Flavanone

Thus differences in the state of oxidation of the central ring give rise to the different classes of flavonoids. But substitutions may also take place in both the A and B rings. Usually the A ring is substituted with hydroxyl groups in the 5 and 7 positions, while in the B ring common substitutions are hydroxyls in the 3′, 4′ and 5′ positions.

One of the most important groups of the flavonoids are the anthocyanins. In plants they occur as glycosides, their aglycones being known as anthocyanidins. Examples of some common anthocyanidins are given below

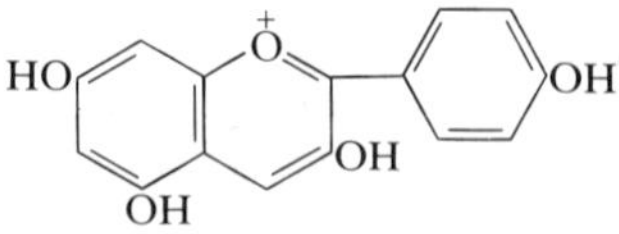

Pelargonidin

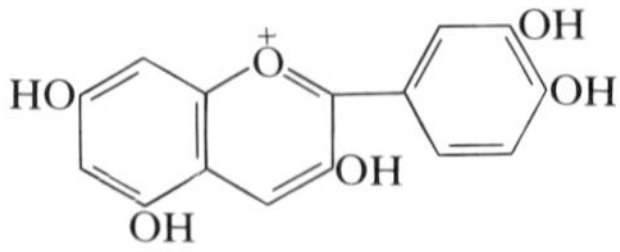

Cyanidin

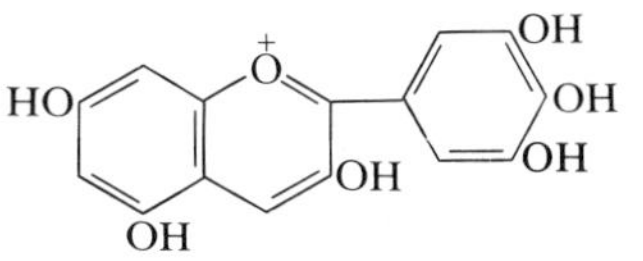

Delphinidin

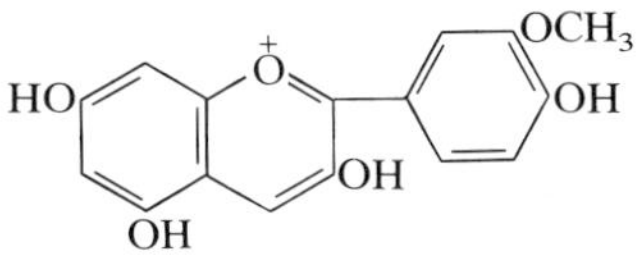

Peonidin

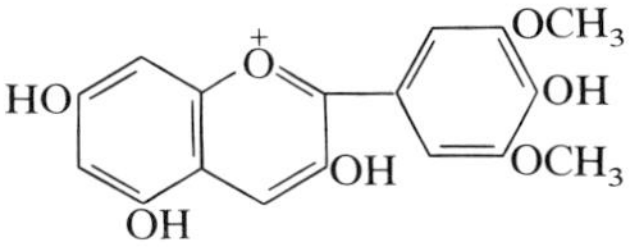

Malvidin

The nett positive charge arising from the structure associated with the oxygen atom in position 1 means that the molecule is basic and can form salts such as chlorides. Consequently the pigments are markedly affected by the pH of the solution in which they are dissolved.

The glycosides are more stable than the aglycones. Glucose is the commonest anthocyanin sugar but galactose and rhamnose also occur. Carbons 3 and 5 are the commonest point of attachment.

Anthocyanins containing adjacent and unsubstituted hydroxyl groups chelate readily with metals such as iron, tin and aluminium, a state of affairs which may cause acute embarrassment to the food processor whose customers are unlikely to appreciate the rather extraordinary colours which can be produced from this reaction in products such as canned fruits and jam. They range from blues to slate greens and may render otherwise excellent products quite unmarketable. For this reason pigmented vegetables such as red cabbage or fruits such as blackberries must be handled using stainless steel equipment and canned in lacquered cans. Some anthocyanins, such as the pelargonadin 3-monoglucoside of strawberries, are also heat-labile degrading to brownish products.

Together with gallic acid, which can be

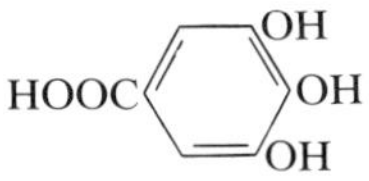

Gallic acid

found esterified through the 3-hydroxyl group of flavanols such as catechin, the flavonoids form the major plant source of polyphenolics. In addition to providing many of the attractive natural colours of fruits and vegetables, polyphenolics are also responsible for undesirable changes such as blackening of cooked potatoes, and the browning of cut surfaces of fruits such as apple, pear, peach and plum. This browning reaction is catalysed in presence of air by a somewhat indeterminate system of enzymes often referred to as 'tyrosinase' or 'polyphenol oxidase'. The reaction can be troublesome in food-processing operations, but it can often be inhibited or temporarily controlled by the use of sulphites, ascorbic acid or even dilute salt solutions. On the other hand, the same reaction is used to advantage in the 'fermentation' of tea, which is not really a fermentation process but one in which the leaves are bruised to allow the polyphenol oxidase in the leaf access to the flavanols present.

This very brief account of the flavonoids can only list a few of the more important properties of this large group of reactive plant substances. They also influence the flavours of fruits and vegetables.

7.11. Flavour of fruits and vegetables

Forty years ago, my professor of organic chemistry, discussing the formation of esters, casually mentioned the possibility that the flavours of fruits and vegetables could be due to the production of these substances by plants. He talked of amyl acetate and the flavour of pears, of allylisothiocyanate as the active principle of oil of mustard, of allyl sulphide as the essential flavouring matter in onions and garlic as well as diacetyl as the substance responsible for the taste of butter. He went on to point out that flavouring substances as well as the perfumes of flowers are present in such tiny quantities that analysis was beyond the limits of the techniques then available.

In the intervening years the analytical methods have become available. My professor's four statements of identity were aimed in the right direction if somewhat off the mark. The flavours of foods are chemically more complex than he would have considered possible.

The human brain perceives flavour through the taste-buds of the tongue, which detect only sourness (pH), sweetness, saltiness and bitterness, although other nerve-endings in the mouth, in responding to temperature or pain (for example the sensations evoked by some types of pepper) add to the overall taste impression. Aroma is detected by the olfactory epithelium, a small area of sensitive skin in the air passages to the rear of the nostrils. Human sensitivity to traces of aromatic substances in inhaled air is extremely high.

Fruits and vegetables may stimulate these senses responsive to sweetness, sourness and bitterness by their content respectively of sugars, organic acids (malic, citric etc.) and flavonoids, the more complex of which are bitter or astringent or both. Examples of the first two groups are too common to repeat here but bitterness is associated, for instance, with the naringin of grapefruit (aglycone naringenin-4′,5,7-trihydroxyflavanone) and hesperidin (aglycone hesperetin-3′,5,7-trihydroxy-4′-methoxyflavanone) which is found in both sweet and bitter oranges. The aglycones are tasteless.

As a result of intensive attacks on the chemistry of the flavour components of fruits and vegetables since the advent of gas-liquid chromatography coupled with mass spectrometry, it is now known that the aroma volatiles of fruits and vegetables include esters, ketones, alcohols and aldehydes. Terpenes are also found in certain instances and in some vegetables complex sulphides and sulphoxides dominate. Much of this work has been motivated by the idea that a complete analysis of the flavour volatiles of a given food would permit the exact replication by flavour chemists of natural flavours, so that, for example, aroma lost in heating processes could be accurately replaced. In the event the situation proves more complex than had been expected. Many fruits and vegetables give very large numbers of different substances on analysis and yet even after the most detailed work, recombination of these in the indicated proportions gives a synthetic mixture which is either a pale shadow of the original or nothing like it at all.

This complexity of the problem has been neatly expressed by S. K. Freeman (*International Science and Technology*, September 1976, p. 72) and I quote

> If 99.999% humulene, which has a woody odour, is mixed with only 0.0001% of the compound ionone, which has an aroma of violets, the mixture will smell only of ionone when sufficiently diluted in air. The explanation lies in the concept of threshold value, which is the minimum quantity of material whose odour can be discerned subjectively. Odour thresholds of substances vary over a wide range. In this case the sensitivity of the nose to ionone (10^{-14} g threshold) is about seven orders of magnitude greater than it is to humulene (10^{-7} g threshold). Obviously only a minuscule amount of ionone will drown out the humulene's odour.
>
> Since the gas chromatography at best can't detect below 10^{-12} g of any material, it is clear that whenever low-threshold odourants are present in a mixture in amounts below the sensitivity of the instrument we will be unable to find them in our analysis even though they are often readily detected by the nose.

Despite this, modern techniques have unravelled very large amounts of information about the chemistry of what are called the ‘flavour volatiles’ of fruits and vegetables. This term refers to the traces of volatile substances either released from the raw fruit or vegetable or the corresponding volatiles released after processing or trapped during a processing stage by an aroma recovery unit. The term is something of a misnomer. For example, over two hundred such substances have been identified in orange juice flavours. By no means all of these are necessarily involved in the flavour impressions one receives on sipping orange juice. While in some cases the flavour involvement of a particular molecular structure may be asserted with confidence, the effort required to investigate all identified substances from a given fruit or vegetable for individual significance is usually beyond the resource of individual workers.

These complications may be academically difficult to manage, but the practical situation is made easier by exploitation of **essential oils**, which can be extracted from most parts (roots, stems, leaves, bark, buds, flowers and fruit) of plants. Some of these have been known and used for fifteen hundred years or more in the production of essences (hence the word essential) used in the manufacture of cosmetics, medicaments and foodstuffs. There are two main groups of isolates.

a) The essential oils which are usually volatiles obtained by steam distillation or by extraction under pressure.

b) The oleoresins which are normally solvent extracts of plant tissues and which may consist in substantial proportion of non-volatile materials. Sometimes oleoresins are distilled to recover a volatile essence.

The distinction between the two groups is practical rather than academic. Oleoresins are applied to non-food uses to perhaps a greater degree than are essential oils. About four hundred essential oils have industrial uses but only about fifty are used in substantial amounts. The oils of anise, almond, caraway, cinnamon, clove leaf, coriander, garlic, ginger, lemon, lime, nutmeg, mace, orange, peppermint, tangerine and thyme are examples of those in major use. Less commonly used are those of cumin, dill, jasmine, juniper, parsley and sandalwood.

In some fruits the aromatic material is present in the juice, but in many cases both in the fruits and in the leaves, the oils are secreted in numerous tiny oil sacs or cells. The essential oils obtained from different parts of the same plant are usually different, as, for example, distilled oil of orange and distilled oil of orange leaf. The chemical substances present fall into two main groups, terpenoids and non-terpenoids.

Terpenes are widely distributed in nature and are generally considered to be derivatives of isoprene, C_5H_8,

$$CH_2{=}\underset{\displaystyle CH_3}{\underset{|}{C}}{-}CH{=}CH_2$$

usually joined end to end, with the following classification of their hydrocarbons.

$C_{10}H_{16}$	monoterpenes
$C_{15}H_{24}$	sesquiterpenes
$C_{20}H_{32}$	diterpenes
$(C_5H_8)_n$	polyterpenes

The naming of the terpenes is non-systematic, but often indicates their most common natural source. The name carries a suffix indication of the main chemical characteristic of the molecule. Thus menthane is a saturated alkane, myrcene has double bonds and is an alkene, graniol is an alcohol with geranial as the corresponding aldehyde and menthone is a ketone, thus

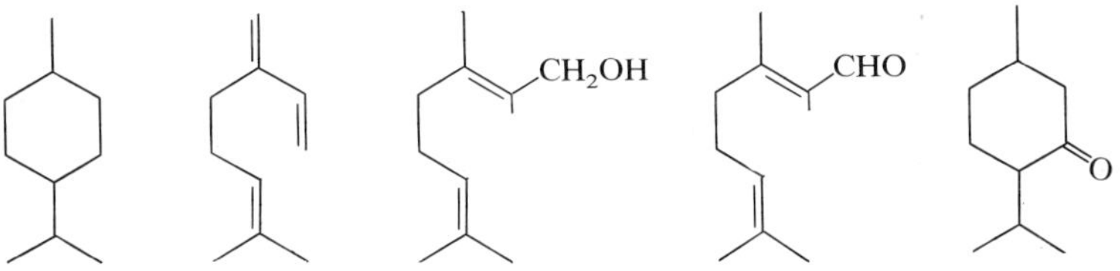

CH_2OH CHO O

menthane myrcene geraniol geranial menthone

Odour characteristics are mainly associated with the monoterpenes. The terpene aldehydes are widely distributed and are especially valued for their aroma characteristics, the aldehyde value of essential oils being regarded as an important index of quality. The sesquiterpenes are less volatile and have weak odour properties. The higher terpenoids are non-volatile and odourless. They form the major fraction of the non-volatile component of the oleoresins.

Non-terpenoid constituents of essential oils include hydrocarbons, alcohols, aldehydes, esters, ethers, ketones and others. They include such common substances as benzyl alcohol (oil of clove), acetic acid (both free and esterified), benzaldehyde (oil of bitter almond), thymol (oil of thyme) coumarin (oil of lavender) and mono-, di- and tri-sulphides (oil of garlic).

Although the essential oils reflect (often in marked degree) the original odour characteristics of their plant of origin, they do not necessarily reflect the changes brought about by cooking or processing. For a discussion of specialised aspects of these, the original literature should be consulted. A general phenomonen does require brief comment here. Most vegetables have characteristic odours but these are mild in the uninjured tissues. Intact cloves of garlic are inoffensive and the pungent aroma only develops when cell rupture allows the enzyme alliinase to attack the odourless compound alliin with the formation of diallyl thiosulphinate, sometimes called allicin.

Alliin is one of a class of compounds termed odour precursors, and they are found in the membership of both the onion family and the brassicas, where again sulphur compounds are principally involved.

Many food-processing operations result in loss of flavour volatiles. Even simple processes such as blanching give rise to some loss of flavour but during concentration and drying processes much of the flavour may be removed to be discarded with the condensate. It is now common practice to recover these lost volatiles by modification of the condensation process. The recovered flavouring constituents may then be returned to the concentrate or dried product, greatly enhancing its quality. Aroma-recovery processes are widely applied in the preparation of instant coffee as well as in the concentration of fruit juices.

7.12. Texture of fruits and vegetables

With vegetables such as celery and lettuce, which are usually eaten raw, as well as with most fruits the texture or crispness is a determining factor in acceptability. Wilting is universally recognised as the enemy of freshness, and great care is taken after harvesting to retain the original texture of the product through distribution channels and right to the point of sale, a particularly difficult task with plants of high moisture content.

Plant cells contain solutes surrounded by a permeable cytoplasmic membrane. Water drawn from the roots moves into the cell solutions increasing the internal pressure so that the cells are blown up and pressed against each other rather like a box full of inflated balloons. The internal pressure is balanced by the elastic walls of the cells, and the resulting equilibrium is referred to as the turgor pressure of the plant structure. If the vapour pressure of the air is lower than that of the cell sap, water will be lost to the exterior and the tissues become limp and flaccid. Crispness may often be maintained by storing fruits and vegetables in a water-vapour-saturated atmosphere but answers to food problems are seldom simple, and such conditions may stimulate mould growth.

The cell structure is also disrupted by the action of heat and by freezing, both giving rise to disorganisation of the cell wall with a consequent loss of turgidity. The cell wall can be conveniently visualised as a framework or trellis of cellulose fibres acting as reinforcing material embedded in and surrounding which is an infilling of hemicelluloses and pectins. The middle lamella, which 'cements' adjoining cells, is discontinuous and the cells do not fit exactly against one another. The spaces at the 'corners' or points of discontinuity are partially interconnected and thus provide access to the external environment for gaseous exchange. These intercellular spaces are relatively large in fruits like the apple which may have up to 25 per cent of its volume as intercellular space compared to about 1 per cent in the case of the potato. Some cells have their structures reinforced by lignin which consists of complex polymers of aromatic units and which is very resistant to chemical and enzymic attack. These structural factors vary from plant to plant in the degree of toughness or tenderness which they impose in the raw state. Their proportions also control the extent of softening brought about by a given cooking or freezing process.

The pectins in the cell walls can form gels under appropriate circumstances and in the presence of acids and sugars. They are therefore of considerable economic importance in processes such as the manufacture of jams and jellies. On the other hand, high pectin levels can be troublesome in fruit juice manufacture as well as in the preparation of certain fruit wines. In these circumstances their interference can often be satisfactorily controlled by induced enzymic breakdown.

7.13. Ripening of fruits

The ripening of fruits is a biochemically dramatic event often involving rapid changes from toughness to tenderness, from bitterness to sweetness and from chlorophyll green to the spectral range of colours of ripe fruit. At the point of ripeness the seeds within the fruit have matured. In contrast, most vegetables, except peas and beans, are eaten long before seeds have set. The circumstances of harvest and storage differ accordingly.

One of the most striking phenomena associated with the ripening of many (although by no means all) fruits is referred to as a climacteric. Respiration rates in fruits are indicative of the rate of compositional change in process. Respiration rates are usually measured as mg CO_2 per kg material per hour. High respiration rates mean high rates of compositional change. Peas for freezing or canning are harvested when somewhat immature because their flavour is preferred at this stage. Harvested shelled peas have respiration rates of several times those of more normal tissue. As a result they quickly lose their sugars and must be processed very quickly after harvesting if quality is to be maintained. Most fleshy fruits, including the tomato, show a sharp rise in respiration rate which coincides with the normal obvious changes of texture, flavour and colour associated with ripening. This respiratory episode in which the rate doubles or even quadruples is often referred to as a climacteric and the fruits which exhibit it are called climacteric fruits. Certain fruits including the pineapple, grape, the citrus group and figs do not exhibit this phenomenon.

Since fruits and vegetables are normally transported considerable distances over varying periods of time between the points of harvest and sale, over-rapid respiration may carry the product past its quality peak and into early senescence before it reaches its final user. The control of respiration rates is therefore of economic importance, and many fruits are harvested in a pre-ripe stage and subjected to controlled ripening conditions thereafter. Respiration rates can be influenced as follows.

1. By lowering the temperture. There are practical limits to this. Tropical fruits are subject to physiological injury if the temperature is

lowered too far, and this is expressed in the accumulation of intermediate metabolites which cause off-flavours. In climacteric fruits, lowering of temperature delays the onset of the climacteric and hence of ripening.

2. By controlling the atmospheric concentrations of oxygen and carbon dioxide. During aerobic respiration, oxygen is absorbed and carbon dioxide released. Thus the storage atmosphere affects respiration rates, increased levels of carbon dioxide or reduced levels of oxygen or both tend to slow the process down. However, there are limits. Too low an oxygen content leads to the accumulation of acetaldehyde and ethyl alcohol in the tissues. Too high a concentration of carbon dioxide gives rise to tissue damage.

3. Ripe fruit placed beside unripe fruit in a closed room hastens ripening, and this has long been known. Unripe fruit placed in rooms warmed by naked coal gas or paraffin stoves ripens more quickly than the same fruit in rooms heated electrically to the same temperatures, and this device has long been used in the induced ripening of bananas, which are cut and shipped green and ripened at their destination. If shipped ripe they are usually overripe or rotten on arrival. The observations are linked by the finding that both ripe fruit and gas jets release traces of ethylene into the atmosphere and ethylene produces the catalytic effect, which is observed in most cases at ethylene levels as low as 1 part per million, although for the commercial use of ethylene to accelerate ripening and the 'degreening' of citrus fruits considerably higher concentrations are recommended. Most plant physiologists now regard ethylene as a hormone and while much experiment and many suggestions have been made as to its role in plant metabolism, the detailed biochemistry of its action remains rather obscure.

Further reference

The Chemical Composition and Nutritive Value of Vegetables, Pyke, M., in *The Nation's Food* (1946) Bacharach, A. L. and Rendle, T. (eds), pp. 108–126, Society of Chemical Industry, London.

Fruits and Vegetables, Charley, Helen, in *Food Theory and Applications* (1972) Paul, Pauline, C. and Palmer, Helen, H. (eds), pp. 251–334, John Wiley, New York.

The Biochemistry of Fruits and Their Products (1970, 1971) Hulme, A. C. (ed.), vol. 1 and 2, Academic Press, London.

The Potato (1966) Burton, W. G. 2nd rev. edn, Vunman and Zonen, Wageningen.

Potato Processing (1959) Talburt, W. F. and Smith, O., Avi, Westport.

Formation and control of Chlorophyll and Glycoalkaloids in Tubers of *Solanum tuberosum* L. and Evaluation of Glycoalkaloid Toxicity, Jadhav, S. J. and Salunkhe, D. K., in *Advances in Food Research* (1975) Chichester, C. O. (ed.), vol. 21, Academic Press, New York.

Manioc in Africa (1959) Jones, W. O., Stanford University Press.
Tropical and Subtropical Agriculture (1961) Ochse, J. J., Soule, M. J., Dijkman, M. J. and Wehlburg, C., vol. 2, Macmillan New York.
Cassava Processing and Storage (1974) Araullo, E. V., Nestel, B. and Campbell, Marilyn (eds), IDRC-031e, International Development Research Centre, Ottawa.
Chronic Cassava Toxicity (1973) Nestel, B. and MacIntyre, R. (eds), IDRC 010-e, International Development Research Centre, Ottawa.
Fruit and Vegetables (1966) Duckworth, R. B., Pergamon Press, Oxford.
Plant Pigments, Flavours and Textures: The Chemistry and Biochemistry of Selected Compounds (1979) Eskin, N. A. M., Academic Press, New York.

8

Et Cetera

8.1. Introduction

In any description of the composition and properties of the principal groups of foodstuffs one is left at the end with a list of topics which lie somewhere between food science and food technology. For example, oils and fats are commodities in the sense that they are marketed on a world-wide basis at prices determined by international demand. But to give them a chapter to themselves would be to distort a situation in which the chemistry of lipids is rightly regarded as the prerogative of the organic chemist, the growing of commercial vegetable oils is the province of the agriculturalist and the extraction and purification of the oils the function of the food or chemical technologist. They nevertheless deserve mention in a book on food science.

Then there are other awkward topics which lie between food technology and food science. 'Browning' reactions fall into this category. Enzymic browing is conveniently treated as an aspect of plant biochemistry. Non-enzymic browning is normally associated with a food processing operation of some sort. Yet because of the superficial similarity of the end-products of these two processes, they are often considered together. Is this reasonable, or should non-enzymic browning not be considered under the heading of processing since it is so often a consequence of a processing operation?

Other topics come to mind. Food quality assessment is a large topic involving physical, chemical and microbiological estimates as well as organoleptic criteria. Then there are the important considerations associated with food hygiene and food toxicology which should be emphasised as key factors in the education of any food scientist or food technologist. Yet all of these topics can best be understood after the student's studies in food processing techniques have made significant progress.

This closing chapter is aimed at giving the student an account of these topics as an overture to his studies of food processing and distribution. It is necessarily, an assemblage of odds and ends not dealt with elsewhere in this volume. As knowledge grows these topics will no doubt find their place in the mainstream of the subject. For the time being it is convenient to introduce them as a link between science and technology.

8.2. Oils and fats

Lipid chemistry has already received a brief discussion under the introduction to nutrition in Chapter 1 (see pages 12–15). The oils and fats of commerce are mixtures of mixed triglycerides with (usually) small amounts of other lipid materials, such as phosphatides, sterol esters, free fatty acids, and waxes, together with fat-soluble pigments such as the carotenoids which may also occur in esterified form.

The reader will also recall that glycerol esters of short-chain fatty acids give softer fats than those of the long-chain acids and that triglycerides containing substantial proportions of unsaturated fatty acids are usually liquid oils at normal ambient temperatures while the more saturated fats are solids. Fats and oils are derived from both animal and vegetable sources and while these have always been present in human food the isolation of these and their industrial processing is, with one or two important exceptions, historically a recent development. Perhaps the two important exceptions are butter, which has been made for at least four thousand years, and olive oil, which has been used as a cooking oil since classical times in the Mediterranean.

The modern oils and fats industry stems in part from the use of these in soap and paint manufacture, which is no direct concern of ours, and in part from the invention of margarine by Mège-Mouriès of Paris in 1869. Both processses were assisted by the rapid population growth in Europe and North America during the 19th century which stimulated demand for all foodstuffs in an unprecedented fashion. The world search for the treasure of unexploited edible oil resources encouraged whaling, the killing and processing of farm animals for fat rather than meat in some parts of the world, and the exploitation of tropical and semi-tropical crops ranging from coconut and oil palms to ground-nuts and soya beans.

The over-exploitation of whales has reduced this source to a trivial proportion of the world supply and rapid growth in world demand for meat has prevented the waste of killing cattle solely for fat. The balance has now been struck by increasing the supplies of vegetable oils, a move of considerable economic benefit to the countries where climate and soil favour their production. Table 8.1 lists some of the more important oil seeds.

The figures given in the table for melting points are rough approximations since natural oils and fats seldom have a true melting point. Being mixtures of triglycerides, of which each is different in molecular weight and often also in degree of unsaturation, natural oils have zones of melting as one triglyceride after another loses its crystalline form.

If oils and fats add interest to cooking (many flavouring substances are soluble in them and they therefore tend to 'hold' or retain cooking aromas), they add problems to the lot of the food scientist. They are subject to two main types of change both of which give rise to the group of off-flavours often described as 'rancidity' in everyday speech.

In the first type, hydrolysis of ester bonds leads to the formation of free

Table 8.1. Some commercial oil seeds

Plant	Characteristics	Melting point
Coconut (*Cocos nucifera*)	high levels of short-chained saturated fatty acids	25°C
Oil palm (*Elaeis guineensis*)	oil from both fruit and kernel	
	palm oil	40
	palm kernel oil	29
Soya bean (*Glycine max*)	high linoleic acid	liquid (−12)
Ground-nut (*Arachis hypogaea*)	high oleic/linoleic acid	liquid (+2)
Olive (*Olea europaea*)	high oleic acid	liquid (−2)
Maize (*Zea* spp.)	high oleic/linoleic	liquid (−12)
Cottonseed (*Gossipium* spp.)	high linoleic acid	liquid (−3)

fatty acids or soaps some which have unpleasant flavours. Triglycerides are easily hydrolysed by heating in alkaline conditions but since most foodstuffs are neutral or slightly acid in reaction, hydrolysis of this sort is relatively uncommon although by no means unknown. More common is hydrolysis or partial hydrolysis by lipase enzymes. Some of these show selectivity towards the positions of the ester bonds. For example, pancreatic lipase has a 'preference' for shorter chain fatty acids and for the 1,3 positions on the glycerol molecule. As the ester bonds are broken, intermediate di- and monoglycerides are formed along with free glycerol. Some lipases continue to react slowly at low water activities and their action has been noted even in dehydrated foods. Hydrolytic rancidity is quite common in cream, nuts, butter and in some types of biscuits, and it occurs most rapidly in emulsion systems where the large lipid-water interface provides easy access of the enzyme to its substrate, imparting a harsh acrid taste to the food. Lipases may be natural constituents of foods or they may be introduced as exo-enzymes from moulds growing on them. The characteristic flavour of blue cheeses is due (in part at least) to short-chain fatty acids produced in this way. Free fatty acids are often produced in vegetable oils during harvesting and extraction. They are removed as soaps in the refining process but a high free fatty acid level reduces the commercial value of the oil, and modern harvesting and extraction methods are aimed at keeping this figure as low as possible.

A much more frequent and serious type of rancidity results from oxidation. This may occur spontaneously when lipids are exposed to air and the reactions involved are collectively referred to as autoxidation. Autoxidation reactions are catalysed by traces of heavy metals and haematin compounds such as haemoglobin and the cytochromes. Characteristically, fats exposed to air react only very slowly with oxygen during the early stages of the reaction. However, the end of this induction period is followed by an autocatalytic reaction which may proceed very rapidly indeed.

The central reaction mechanism was first proposed by Farmer and after more than thirty years has now found a degree of general acceptance.

The reaction begins with a molecule of lipid RH being sufficiently activated by heat or light or the presence of a heavy metal catalyst to decompose into the unstable free radicals $R^{\cdot}$ and $H^{\cdot}$. Most of these will quickly disappear as H_2, RR, RH etc. but in the presence of oxygen $R^{\cdot}$ may form a peroxide radical $ROO^{\cdot}$, which then reacts with a fresh molecule of lipid

$$ROO^{\cdot} + RH \rightarrow R^{\cdot} + ROOH,$$

giving rise to the new radical through which the chain is propagated. The reaction is terminated when free radicals combine with each other or with free radical inactivators to form stable products. The hydroperoxides formed are relatively unstable and decompose and degenerate to form alcohols, aldehydes and ketones which contribute to the off-flavours generated.

These degradation reactions may be associated also with polymerisation. The highly unsaturated oils used in paint manufacture polymerise in this way as the paint dries and are often referred to as drying oils. In foods autoxidation reactions are undesirable and unless controlled may result in substantial losses. Exclusion of air is the first line of defence but a number of chemical antioxidants are highly effective in inhibiting autoxidation. These are free radical acceptors which, by absorbing the radicals as they are formed, break the chain sequence. The naturally occurring tocopherols, of which vitamin E is an example, act in this way. Certain synthetic phenolics are effective antioxidants. Of these, butylated hydroxyanisole (BHA), butylated hydroxytoluene (BHT) and propyl gallate (PG) are recognised as safe food additives in many countries. They are used at levels of up to 200 parts/million in the fat or oil being protected.

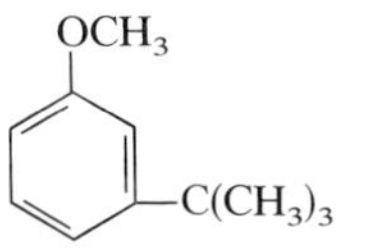

OCH_3 / $C(CH_3)_3$

Butylated hydroxyanisole

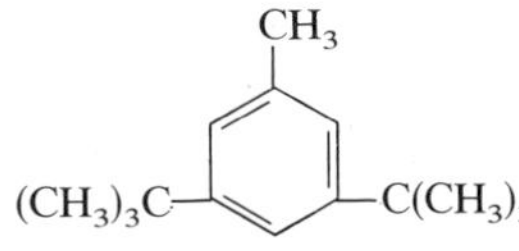

CH_3 / $(CH_3)_3C$ / $C(CH_3)_3$

Butylated hydroxytoluene

$C(=O)$—O—C_3H_7 / HO / OH / OH

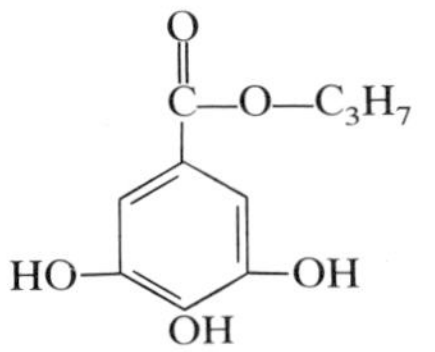

Propyl gallate

Their action as chain breakers can be illustrated as follows

$$ROO^{\cdot} + AH \rightarrow ROOH + A^{\cdot}$$

$$A^{\cdot} + A^{\cdot} \rightarrow A-A$$

or

$$A^{\cdot} + ROO^{\cdot} \rightarrow ROOA.$$

Peroxidation of fats may also be catalysed by the action of the enzyme lipoxygenase. This enzyme is principally found in legumes, soya beans being a noted source, but it also occurs in other vegetable materials such as oil seeds and cereal grains. While autoxidation may occur with any unsaturated fat, the action of lypoxygenase is specific to the structure (*cis*, *cis*)

$$R_1\text{—CH=CH—CH}_2\text{—CH=CH—}R_2$$

with linoleic, linolenic and arachidonic acids and their esters as the significant substrates. It catalyses the initial reaction

$$R_1\text{—CH=CH—CH}_2\text{—CH=CH—}R_2 + O_2 \longrightarrow R_1\text{—}\underset{cis}{\text{CH=CH}}\text{—}\underset{trans}{\text{CH=CH}}\text{—}\overset{\overset{\text{OH}}{|}}{\underset{}{\overset{\text{O}}{\overset{|}{\text{CH}}}}}\text{—}R_2$$

which may be followed by similar degradation reactions to those of autoxidation.

If carotenoid pigments are present in oxidising oils and fats they too are oxidised in coupled reactions which are of nutritional significance in terms of losses of vitamin A activity. This reaction takes place both during autoxidation and enzyme-catalysed lipid oxidation. The latter reaction has found some useful industrial application in baking technology by bleaching flour pigments (carotenoids devoid of vitamin A activity) during the dough-mixing process in presence of a small addition of soya flour.

As well as giving rise to off-flavours, there have been suggestions by some workers that the oxidation products of oils and fats may be harmful to human health if consumed in quantity. The evidence is somewhat contradictory and the current view is they are probably harmless in the quantities likely to be consumed in the western world. Although the flavour of oxidised fat is usually regarded as rather unpleasant and often as offensive by the Caucasian races, some eastern peoples seem to find it either acceptable or even attractive. Assuming that traditional products so affected have been consumed in these countries for generations, it is likely that any acutely toxic effects would by now have been observed.

Some of the characteristic spoilage micro-organisms found in food produce fat-hydrolysing or oxidising enzymes. For example *Strep. cremoris* and *Oidium lactis* are repoted to cause rancidity in cream while *Candida lipolytica* is associated with similar spoilage in margarine.

Cream, margarine, butter and mayonnaise are all examples of emulsions. These may consist of oil droplets in suspension in a continuous aqueous phase or aqueous droplets in suspension in a continuous fat phase, cream and mayonnaise being examples of the former and butter and margarine of the latter. Triglycerides, being essentially non-polar, quickly separate when suspended in water. However, if a substance is introduced which combines polar with non-polar properties, suspensions of oil-in-water or water-in-oil can be stabilised. For example, the substance glyceryl monostearate

$$CH_3—(CH_2)_{16}—\overset{\overset{\displaystyle O}{\|}}{C}—O—CH_2—CH(OH)—CH_2(OH)$$

is a combination of the non-polar C_{18} fatty acid chain with the two polar hydroxyl groups of glycerol. When a triglyceride is suspended in water in the presence of such a molecule, the non-polar end 'dissolves' in the oil droplets but the polar group is unable to do so, thus forming a kind of membrane or surface layer surrounding each droplet with the polar groups protruding into the aqueous medium and the non-polar chains oriented towards the droplet's centre. Substances behaving in this way are known as emulsifying agents, and when present in appropriate concentrations prevent the coalescence of adjoining droplets. If the initial size of the droplets is large, oil being lighter than water, they tend to rise to the surface of the system forming a cream in the same way that cream rises to the top of milk. Emulsion separation of this kind can be reduced or entirely prevented by reducing the particle size by mechanical means or by increasing the viscosity of the continuous phase. Substances which perform the latter function are known as stabilising agents. The formation of food emulsions and their subsequent stability is a matter of considerable industrial importance.

The substance sold commercially as glyceryl monostearate is usually a mixture of mono- and diglycerides. In fact a large number of substances are commercially available and legally permitted in most countries as emulsifying and stabilising agents. Some of these are normal constituents

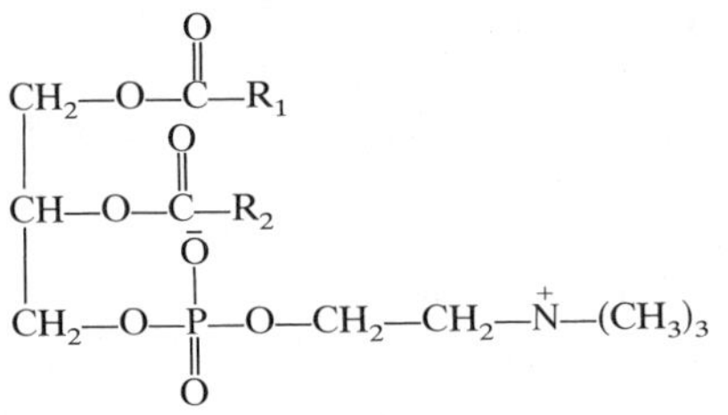

$$\begin{array}{l} CH_2—O—\overset{\overset{\displaystyle O}{\|}}{C}—R_1 \\ | \\ CH—O—\overset{\overset{\displaystyle O}{\|}}{C}—R_2 \\ |\qquad\quad \overset{-}{O} \\ |\qquad\quad | \\ CH_2—O—\underset{\underset{\displaystyle O}{\|}}{P}—O—CH_2—CH_2—\overset{+}{N}—(CH_3)_3 \end{array}$$

Lecithin, R_1 and R_2 being fatty acid residues

of foodstuffs. For example, the well-known emulsifying properties of egg-yolk are mainly due to its content of lecithins. Others are chemically modified natural products or of synthetic origin. Many of the substances used as stabilisers are natural gums such as acacia, locust bean and tragacanth. As with emulsifiers, others such as the methyl and ethyl celluloses are chemical modifications of readily available natural products.

The extraction of edible oils from their parent plants, their purification, hydrogenation and manufacture into products for direct use or for incorporation into a wide range of foodstuffs has given rise to large industries employing sophisticated technologies.

8.3. Non-enzymic browning reactions

8.3.1. The Maillard reaction

In 1912 a certain M. L.-C. Maillard published a short note on the production of what he called melanoid pigments by reacting glycocoll (the old name for the amino acid glycine) with glucose. His paper is worth rereading today (*Comptes Rendus Hebdomadaires des Seances de l'Academie des Sciences, France* 1912, *154*, 66–68). He describes one experiment as follows

> One introduces 0.5 g of glycocoll, 2 g of glucose, 2 ml of water and 14.6 ml of oxygen under a bell-jar over mercury. After six hours at 100°, one recovers 12.6 ml of oxygen (2 ml only having disappeared) and 22.9 ml of carbon dioxide. Thus this gas could not have derived its oxygen from the atmosphere.
>
> Using an arrangement for absorbing the gaseous products, 0.4993 g of glycocoll and 2.0044 g of glucose had produced in seven hours 0.1048 g of CO_2. After drying at room temperature in the open, the residue represents a loss of 0.6227 g of which 0.5179 is something other than CO_2. Analysis of this residue shows that this difference is water...

He goes on to explain that he repeated the experiment with other sugars and other amino acids. While the reaction rate varied, the end results appeared to produce essentially the same brown pigments. He also noted that the reaction was slow with lactose and maltose and very slow indeed, if at all, with sucrose. He further suggested that these reactions were of general importance in plant physiology, in agronomy and perhaps also in the geology of carboniferous minerals.

As he himself pointed out in a later paper (*C. R. Hebd. Seances Acad. Sci.*, 1912, *155*, 1554–1556), A R. Ling had preceded him in noting this reaction as a source of the colouration developing in malt at the completion of the malting process (*J. Inst. Brew.* 1908, *14*, 514) and demonstrating it *in vitro* with asparagine and glucose. However, Ling had merely noted the reaction in the narrow context of a brewing process, had not observed the losses of water and carbon dioxide, had not noted the presence of nitrogen in the reaction products nor had he associated it with the naturally occurring melanoid pigments. Despite Ling's claim for priority, the reaction is now universally known as the Maillard reaction.

The principal characteristics of the reaction can be summarised as follows.

1. It is a general reaction taking place between reducing sugars and substances containing free amino groups, such as amino acids and proteins.
2. Carbon dioxide and water are evolved during the course of the reaction.
3. The brown end-products of the reaction are mixtures of polymerised materials of uncertain composition.
4. The reaction is of widespread occurrence in foods. In some instances (the browning of bread crusts, for example) it is desirable; in others (browning in dried milk, for example) it is deleterious and gives rise to commercial losses.
5. The reaction can only be completely prevented in practice by removal of one of the reactants.
6. The reaction rate can be reduced (a) by lowering the temperature of storage; (b) by optimising the water activity (a_w) (the reaction rate is lowest at very low and very high water activities and reaches a maximum velocity at some intermediate value); (c) by the use of chemical inhibitors of which sulphur dioxide is the most important; (d) by lowering the pH of the system.
7. In addition to modifying food color, browning reactions are believed to contribute to certain types of odours and flavours, browning reaction products of valine being associated with rye-bread odours, of aspartic acid with caramel odours, leucine with that of sweet chocolate, and many more.
8. The initial stages of reaction do not require the presence of oxygen (see, in contrast, ascorbic acid browning as described in the next section), although the final stages of polymerisation of the melanoidins formed is affected by the availability of oxygen.
9. The brown pigments produced vary from being readily soluble in water to being virtually insoluble. The formation of soluble brown pigments precedes the formation of the highly polymerised insoluble material. The term melanoidin is reserved for this insoluble material and should not be applied to the soluble pigments produced in the earlier stages of the reaction.

Despite much investigation the chemistry of the browning reaction is imperfectly understood. After a fairly straightforward start, which can be summarised as shown in the reaction scheme.

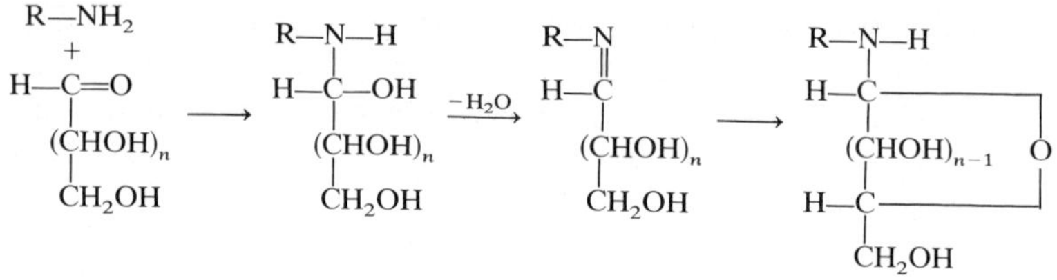

$$\begin{array}{c} R\text{—}NH_2 \\ + \\ H\text{—}C\text{=}O \\ | \\ (CHOH)_n \\ | \\ CH_2OH \end{array} \longrightarrow \begin{array}{c} R\text{—}N\text{—}H \\ | \\ H\text{—}C\text{—}OH \\ | \\ (CHOH)_n \\ | \\ CH_2OH \end{array} \xrightarrow{-H_2O} \begin{array}{c} R\text{—}N \\ \| \\ H\text{—}C \\ | \\ (CHOH)_n \\ | \\ CH_2OH \end{array} \longrightarrow \begin{array}{c} R\text{—}N\text{—}H \\ | \\ H\text{—}C\text{—} \\ | \\ (CHOH)_{n-1}\;\;O \\ | \\ H\text{—}C\text{—} \\ | \\ CH_2OH \end{array}$$

Aldose — Schiff base — Aldosylamine

Formation of substituted aldosylamine (step 1)

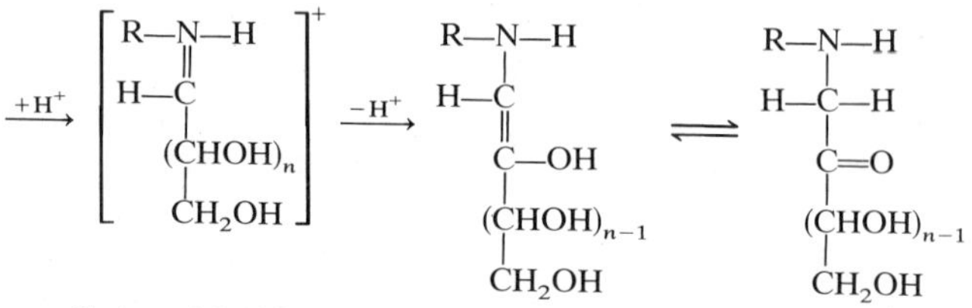

$$\xrightarrow{+H^+} \left[\begin{array}{c} R\text{—}N\text{—}H \\ \| \\ H\text{—}C \\ | \\ (CHOH)_n \\ | \\ CH_2OH \end{array}\right]^+ \xrightarrow{-H^+} \begin{array}{c} R\text{—}N\text{—}H \\ | \\ H\text{—}C \\ \| \\ C\text{—}OH \\ | \\ (CHOH)_{n-1} \\ | \\ CH_2OH \end{array} \rightleftharpoons \begin{array}{c} R\text{—}N\text{—}H \\ | \\ H\text{—}C\text{—}H \\ | \\ C\text{=}O \\ | \\ (CHOH)_{n-1} \\ | \\ CH_2OH \end{array}$$

Cation of Schiff base — Enolic form — Ketonic form

Amadori rearrangement (step 2)

The Amadori compounds are the key intermediates in the autocatalytic reactions which follow. From a study of the reaction products produced in model systems it seems that in a sequence of reactions amine is eliminated with the formation of structures of the type

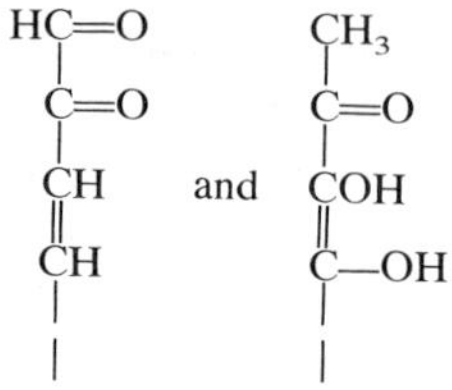

$$\begin{array}{c} HC\text{=}O \\ | \\ C\text{=}O \\ | \\ CH \\ \| \\ CH \\ | \\ | \\ \text{I} \end{array} \quad \text{and} \quad \begin{array}{c} CH_3 \\ | \\ C\text{=}O \\ | \\ COH \\ \| \\ C\text{—}OH \\ | \\ | \\ \text{II} \end{array}$$

Each of these structures may now react with amine by the Strecker degradation mechanism which can be shown in generalised form as

$$R_1\text{—}\overset{O}{\overset{\|}{C}}\text{—}\overset{O}{\overset{\|}{C}}\text{—}R_2 + R_3\text{—}CH(NH_2)COOH \longrightarrow R_1\text{—}C(NH_2)\text{=}C(OH)\text{—}R_2 + R_3CHO + CO_2$$

isotopic labelling having demonstrated that most of the CO_2 originates from the amino acid. The ultimate brown pigments contain nitrogen some of which has almost certainly been incorporated by this reaction.

However, compounds of type I may also give rise to furfuraldehydes by elimination of H_2O, which may then react with more amine in a further sequence leading to brown pigments. Compounds of type II may produce pyruvaldehyde, diacetyl and hydroxydiacetyl, which may also react with further amine to give melanoidins.

Thus starting with a very simple amino acid/sugar system, a wide range of intermediate products may be produced. Most of these, taken individually, can undergo browning reactions in the presence of free amino groups, indicating their involvement in the overall browning reaction. While model systems of this sort have proved a valuable tool in the study of Maillard reaction mechanisms, the reaction may also take place between any carbonyl and free amino group present in a food. Thus the ε-amino groups of lysine-containing proteins may be involved with attendant loss of biological value. Ketoses as well as aldoses participate. Among the reaction products formed which contribute to food flavours are furfurals, pyrones, pyrazines and a range of carbonyls.

The reaction rate and the degree of polymerisation produced depends on factors such as time, temperature, the water activity of the system, the type of sugar involved and pH.

The final stages of the reaction involve random polymerisation of the carbonyl intermediates with or without the participation of further amino acids, the final reaction products being brown melanoidins of unknown structure.

The Maillard reaction is regarded as deleterious in dried fruit since it gives rise to darkening in apricots, peaches, pears and sultanas. It also causes colour deterioration in products such as dehydrated meat, milk, eggs, fish, fruit juices and vegetables. In contrast browning reactions contribute to the aroma, flavour and colour of many foods such as bread (crusts), breakfast cereals, potato crisps, french fries, potatoes, roasted coffee and malted barley. However, in some of these cases the desired effect or the deteriorative reaction may be only partly due to the Maillard reaction and partly attributable to one or a combination of the other forms of browning reaction described below. Some of the reaction products, the aminohexose reductones, are excellent antioxidants for fats but are toxic to animals. However, the search for non-toxic derivatives is showing promise.

8.3.2. Ascorbic acid browning

Citrus juices are subject to a browning reaction which differs in important essentials from the Maillard reaction. Lemon juice (pH 2.5) and orange juice (pH 3.4) should be protected from the Maillard reaction by their acidity. Furthermore browning in these juices can be inhibited by careful exclusion of oxygen yet oxygen does not seem to be required for the initiation of the Maillard reaction. Studying the reaction in orange juice, Joslyn and Marsh, as long ago as 1935, came to the conclusion that ascorbic acid was the principal reactant and that amino acids played only a minor role if any. A number of other workers have since studied the reaction and Clegg and Morton (*J. Sci. Food Agric.*, 1964, *15*, 878–885 and 1965, *16*, 191–198) have demonstrated the presence of citral, furfural, diacetyl, hydroxymethylfurfural and, with less certainty, α-ketogulonic acid in the reaction products. On the basis of their findings and in those of others working on the same reaction, they have proposed the following reaction mechanism:

Breakdown of ascorbic acid

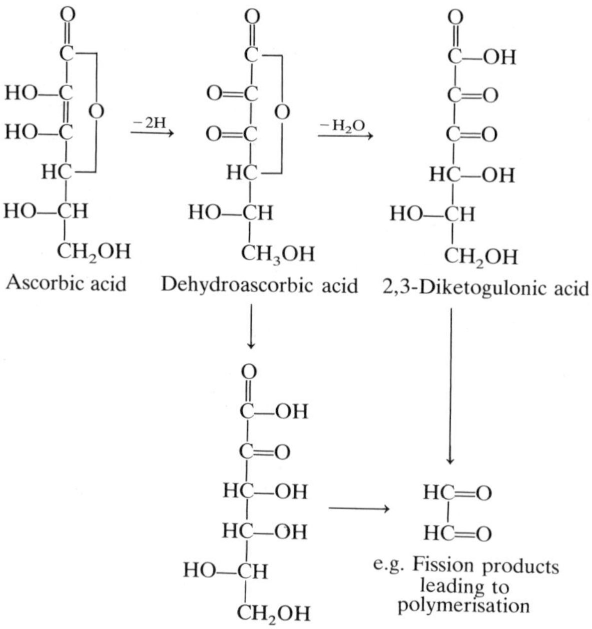

O=C—, HO—C, HO—C, HC—, HO—CH, CH_2OH
Ascorbic acid

$\xrightarrow{-2H}$

O=C—, O=C, O=C, HC—, HO—CH, CH_3OH
Dehydroascorbic acid

$\xrightarrow{-H_2O}$

O=C—OH, C=O, C=O, HC—OH, HO—CH, CH_2OH
2,3-Diketogulonic acid

Dehydroascorbic acid ↓

O=C—OH, C=O, HC—OH, HC—OH, HO—CH, CH_2OH

⟶ (also from 2,3-Diketogulonic acid ↓)

HC=O, HC=O
e.g. Fission products leading to polymerisation

They compared their results on lemon juice with those obtained using model systems and isolated in all twelve carbonyls as reaction products. They confirmed the earlier findings, in that in the first stages of the reaction in model systems the presence of amino acids depressed the reaction rate. On the other hand, the addition of carbonyls to the reaction system, especially unsaturated carbonyls, such as methyl vinyl ketone and crotonaldehyde, markedly accelerated browning in model systems. They also showed that the presence of citric acid was important to the reaction rate.

As far as present knowledge goes, sulphur dioxide appears to help reduce the rate of ascorbic acid browning but the most effective way to retard it in commercial products is the careful exclusion of oxygen from the product or products.

8.3.3. Lipid oxidation and browning

One of the main requirements for browning reactions is the presence of reducing groups such as aldehydes and ketones. Now it is well known that carbonyls are produced during the auto-oxidation of unsaturated fatty acids. One of the main products of oxidation is malonaldehyde and this can react with amino acids to form the carbonyl amino compounds which are the precursors of many browning reactions. The reaction is of some nutritional importance in that it may render unavailable limiting amino acids such as lysine. This has been shown to occur, for example, during the processing of herring meal. It seems that amino acids bound in the browning reaction are not hydrolysed in the digestive tract and some are destroyed in the Stecker degradation. Lysine, it seems, is very susceptible to loss through browning and since it is the limiting amino acid in many foods the effect of browning on the biological value of protein may sometimes be of a substantial nature.

8.3.4. Caramelisation

When sugars are heated above their melting points brown pigments, known as caramels, are developed. Their chemical composition is complex and appears to vary with the heating conditions employed. The reaction occurs under both acid and alkaline conditions and the reaction products are commercially available as viscous black syrups which give brown or red-brown solutions when suitably diluted with water. Commercial caramels are used for colouring soft drinks, beers, spirits, bakery and grocery products. For example, whisky is water white when distilled and thereafter is matured for several years in oak barrels. When withdrawn from the barrels for blending it may be almost as colourless as from the still, or the colour may range from the palest of straw to a light brown, depending on the pigments present in the wood in which the distillate was matured. Commercial whiskies sold under brand names are obtained by blending different distillates to achieve the characteristic flavour desired. This results in a product of variable colour, and were steps not taken to standardise this, consumers might well think that colour variations reflected variations in quality. The colour of each blend is therefore standardised by adding small amounts of caramel solutions before the product is finally sent to the bottling line. Caramelised sugars are regarded as harmless food additives in most countries and are used quite widely as colouring agents. Phosphates, citrates, malates, fumarates and tartarates catalyse the caramelisation reaction.

Apart from producing useful colours, the pyrolysis of sugars, if accurately controlled, gives rise to attractive flavours, a fact which is exploited in the confectionery industry.

The chemical composition of caramels is extremely complex and is imperfectly understood. As a reflection of their complexity J. S. Fagerson (*J. Agric. Food Chem.*, 1969, *17*, 747) has listed 96 compounds which have been isolated from heated glucose, including lactones, aldehydes, ketones, acids, furans, alcohols, and even aromatics. Since most of these are reactive substances, and since glucose in real foodstuffs is always surrounded by other materials which are themselves reactive, the range of possible chemical change is very large.

The task of manufacturing commercial caramel as colouring agents is difficult but not as daunting as it might seem. 'Black Jack', which is the old trade name for caramel colour, is prepared by controlled pyrolysis of sugars in the presence of the catalysts mentioned previously in grades designed for specific applications and to avoid the possibility of precipitation or partial precipitation of the pigment on prolonged standing or storage of the coloured product. It is normally advantageous to seek the advice of the caramel manufacturer on the grade to be used for specific applications.

8.4. Food hygiene and food safety

8.4.1. Introduction

From its beginning humankind has been subject to danger from the presence of toxic substances naturally occurring in the plants or animals used as food, and to further danger from the possibility that these may transmit infection to him through contamination by micro-organisms or infestation by parasites. Natural selection is a hard taskmaster, and our race has long been adapted to its environment, having learned by experience to avoid the hazards presented, for example, by acutely toxic plants. In this process man's metabolic system has developed the capability of dealing effortlessly and harmlessly with the traces of toxic materials which occur in many of our everyday foodstuffs. Likewise the human immunological system has evolved in such a way as to give protection against the moderate ingestion of the food-borne pathogens which occur widespread in our environment. For most of its history our race has concerned itself with hunting or agriculture, activities which encourage dispersion and hinder concentration of populations. Under these circumstances, when contamination by toxins such as ergot or infection by food poisoning organisms did take place, their unpleasant or fatal effects tended to be localised.

Urbanisation has now reversed the 10 per cent town dwellers and 90 per cent rural workers of one hundred and fifty years ago to a situation where, in developed countries at least, most of the population is urbanised. In, historically speaking, a short time period, the balance of hazard has also been reversed and the possibility of large-scale disaster through food poisoning has been correspondingly increased. The fact that large-scale disasters have not taken place is due mainly to the sensible application of scientific knowledge of food safety and food hygiene.

8.4.2. Food poisoning

Certain diseases are transmissible from animal to man, bovine tuberculosis (with milk as the source of human infection) and brucellosis being noted examples. Others may be transmitted from man to man via food or water. These include enteric fevers such as typhoid and paratyphoid, cholera (predominantly water-borne), infective hepatitis (from water and food such as shellfish), bacillary dysentery, amoebic dysentery and similar infections.

In addition there is a group of food-poisoning organisms which are widespread in nature and whose most damaging attacks occur after the organism has grown on or within a food, using the food as its substrate. The illness consequent upon the consumption of such food may be due to infection (the invasion of the tissues of the host by the organism) or to intoxication by toxins excreted into the food by the organisms growing in it. The onset of symptoms from infective food poisoning occurs six to forty-eight hours after consumption while those from toxins develop more rapidly, the first symptoms sometimes appearing within an hour of eating the infected food and almost always within twelve hours. An exception is the case of *Clostridium botulinum*, the organism responsible for botulism. Botulism is the often fatal illness which follows the ingestion of the toxin elaborated by this organism, and whose first symptoms seldom appear in less than twelve hours after consuming the infected food.

The duration of the illness is normally one to seven days in the case of infective food poisoning and six to twenty-four hours with toxin poisoning. Again botulism is an exception. If it does not prove fatal (and the world average mortality rate is about 31% of all recorded cases), recovery is slow and convalescence may take six months or more.

The proper study of these organisms is the province of the professional food microbiologist. On the other hand, the consequences of complete ignorance on the part of any person concerned with the practical handling of food are serious and a brief outline is therefore included here, as an *aide-mémoire* to hazard.

The *Salmonella* group is one of the commonest sources of food poisoning. It is a very large group in which more than one thousand serologically distinct types are recognised. The serotyping is diagnostically important in that it is essential to identify the serotype to establish a firm link between the outbreak and its source in food. The organisms reach food either directly or indirectly from human or animal excreta or from sewage-polluted water. In kitchen or factory the organisms may be transferred to food by infected working surfaces, from utensils, machines and also from human hands. A chance contamination by a small number of the bacilli may not cause illness and it is likely that most of us are mildly infected in this way from time to time. However, even a small initial infection becomes large when the infected food is allowed to stand for a period of a few hours in warm conditions before it is eaten. This is allowed to happen all to often not only in the domestic kitchen but also in catering operations in hotels, restaurants and even in hospital kitchens. It is often from such communal feeding situations that the larger outbreaks originate. Illness normally develops between six and thirty-six hours after eating the infected material, and the food may give no clue by taste, smell or appearance of the heavy bacterial load with which it is infected. The usual symptoms include acute diarrhoea and vomiting, prostration, muscular pain, fever and headache. The illness usually clears up after a few days, but can be fatal to infants or elderly or sick people, twenty to forty fatal cases being recorded annually for example amongst the fifty million people living in England and Wales, where salmonella infections account for about three out of every four recorded cases of food poisoning.

Food poisoning also arises from certain strains of *Staphylococcus aureus*. The skin and nostrils often harbour staphylococci and they may also be spread from respiratory infections, suppurative lesions such as infected cuts, boils and pimples or by coughing and sneezing when suffering from a respiratory infection. When the organism grows in infected food an enterotoxin is produced which is more heat resistant than the organism itself and it is this toxin which produces the illness. Since intoxication is involved, the symptoms (vomiting, diarrhoea, abdominal pain and cramps) usually appear within four to six hours after ingestion but recovery is rapid.

Clostridium perfringens (C. Welchii) is commonly found in excreta from humans and animals and in poultry and raw meats. Its spores are resistant to heat and may germinate and multiply rapidly during long slow cooling after cooking. The poisoning symptoms include abdominal pain, diarrhoea and nausea but vomiting is relatively uncommon. Poisoning follows the ingestion of large numbers of the organisms and the illness is due to the production of an enterotoxin in the intestine. The symptoms usually clear up in about forty-eight hours. Next to *Salmonella* it is the most common cause of food poisoning in the United Kingdom, in an average year being responsible for about fifteen per cent of all cases.

Escherichia coli is a normal inhabitant of the intestinal tract of man. Despite this, some strains are enteropathogenic, especially to infants, and give rise to acute diarrhoeal enteritis with occasional fatalities in hospitals and maternity homes. Adult diarrhoea is also attributable to certain serotypes, and attacks of food poisoning when travelling are sometimes attributable to these.

Vibrio parahaemolyticus is associated with seafoods, especially in warm weather. It is said to be responsible for about half of the incidence of food poisoning in Japan where such foods are a major constituent of local diet. However, there are reports of outbreaks from many countries including the United Kingdom, the United States and Australia. It is an infective illness, the normal incubation period being twelve to eighteen hours and an attack lasts for about two to five days. The symptoms include the usual diarrhoea with vomiting, fever and dehydration. Incidents are associated with the consumption of both raw and cooked seafoods such as shellfish, lobster, shrimps, prawns and crab. Imports of frozen seafoods from eastern countries are sometimes contaminated with the organism, and outbreaks of poisoning have occurred after dinner parties and banquets when prawn cocktail has been on the menu. However, compared with outbreaks associated with *Salmonella*, the rate of incidence is small.

Clostridium botulinum has already been mentioned, and it has received an attention which is out of proportion to its epidemological significance. There are two reasons for this. The first is the extreme seriousness of the consequences when an outbreak does occur, the fatality rate (about 31 per cent) being very much higher than that associated with any other form of food poisoning. The second is the unusual heat resistance of its spores which, in the classical study of Esty and Meyer in 1922, had to be heated for 330 minutes at the boiling point of water in phosphate buffer at pH 7 to ensure their complete destruction, while 4 minutes were required at a temperature of 120°C. Since the organism is an anaerobe, it can grow and produce toxin in insufficiently processed canned or bottled food. It does not develop below pH 4.5. The dangers associated with this organism are such that all low-acid foods (i.e. of pH 4.5) must be sterilised under conditions of time and temperature which ensure the destruction of *C. botulinum* spores.

Fortunately botulism is a relatively rare disease and antitoxins may save life if administered during the early stages. On the other hand, diagnosis of the illness is difficult at these early stages, the only certain method being based on the protection of test animals by antisera. Seven serotypes are identifiable (labelled A to G) but only four, A, B, E and F are known to affect man. The toxins produced by the different serotypes vary somewhat in their resistance to heat but, unlike their parent spores, are readily destroyed by the temperatures normally encountered in cooking. Were it not for this fact, botulism would be a commoner disease than it actually is.

C. botulinum is widely distributed in nature and has been isolated from virgin and cultivated soils in many countries including the United Kingdom, other European countries, the United States, Russia and India. It is also found in the muddy deposits in some inland seas such as the Baltic and from marine sediment off the Pacific coast of America.

C. botulinum infects animals as well as man and is responsible, for example, for the disease known as limberneck in fowls, as well as illnesses in cattle, horses, mink and wild birds. Dogs, cats and pigs are relatively resistant.

Bacillus cereus spores are frequently isolated from cereals and cereal products. The organism is an aerobe which develops and produces toxins in warm storage conditions from spores which have survived cooking. Poisoning outbreaks have been reported from boiled and fried rice, from cornflour and from similar types of products. The incubation period is usually short (one to sixteen hours) and the symptoms are similar to those of staphylococcal poisoning, with acute vomiting and diarrhoea. *Bacillus subtilis* and other aerobic spore-forming bacilli have been implicated in food-poisoning outbreaks with symptoms resembling those of *B. cereus.*

A number of other bacteria, especially streptococci have been implicated in food poisoning outbreaks from time to time but their significance seems minor in relation to the frequency of occurrence of the other forms already mentioned.

Some fungi produce mycotoxins. Poisoning from the consumption of certain species of mushrooms has been known since ancient times. 'Saint Anthony's Fire' or, as it is now known, 'ergotism' is a syndrome known since the middle ages. Epidemics are due to the consumption of grain infested with the mould *Claviceps purpurea,* which elaborates a number of toxic alkaloids related to lysergic acid. Apart from these two examples, mould-damaged crops were for long regarded as acceptable for animal feeds and there are many reports attesting to the absence of harmful effects from their use. Indeed until comparatively recently, moulds were regarded as harmless to man by food scientists and their presence in food was avoided, not on grounds of safety but because they often produced undesirable tastes and spoiled the appearance of products on which they developed. Occasionally their growth was deemed desirable, for example, in certain cheeses and in the manufacture of soya sauce.

In 1960, large numbers of turkeys died in England for unexplained reasons. The mysterious illness soon spread to chickens, ducklings and pigs. The large losses led to intensive investigations extending over feed materials supplied from many countries which established that the causative toxin was elaborated by the growth of the fungus *Aspergillus flavus* in the infected material.

Aided by the early observation that the toxic agents fluoresce, the toxic agents were isolated and their structures assigned by 1964. A number of aflatoxins (as they were called) are now known as variants on one or another of the following structures

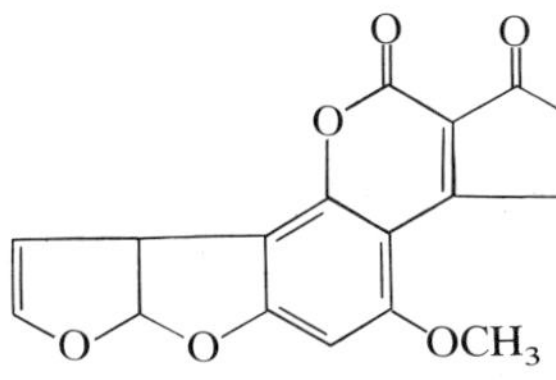

B_1 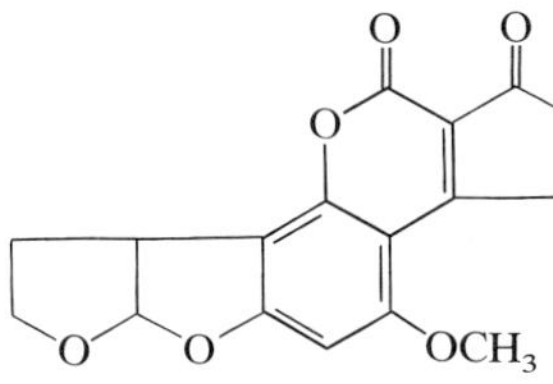 B_2

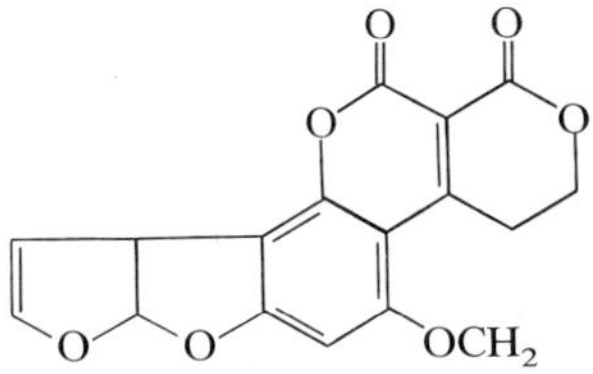

G_1 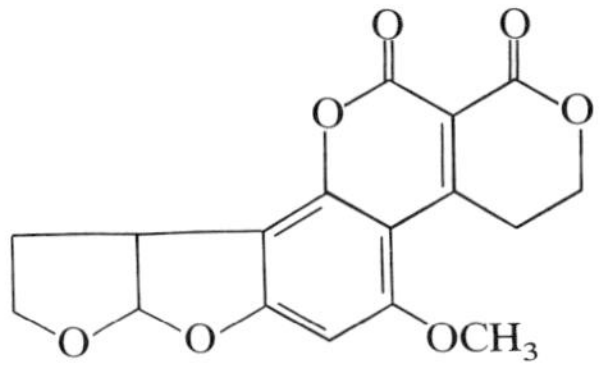 G_2

In addition to demonstrating acute toxic symptoms, aflatoxins are now known to be potent carcinogens. As far as is at present known the main significance of aflatoxins to man is their influence on the health of his domestic animals. However, the possible association of the consumption of food contaminated with these substances in, for example, carcinoma of the human liver cannot be excluded and there is some epidemiological evidence which can be cited in support of this supposition.

However, ergotism apart, the only form of mycotoxicosis which is known to have any serious effect on human populations is alimentary toxic aleukia (ATA) and its incidence has so far has been restricted to certain districts of the U.S.S.R. It is an often fatal disease associated with the consumption of overwintered grain harvested in the following spring. The main vector appears to be millet, but other cereals including wheat and oats have also been implicated. The elaboration of the toxic substances from *Fusarium* and *Cladosporium* species seems to be associated with the metabolism of the moulds when developing slowly in infected grain at low winter temperatures. Toxin production is not limited to these two species, although they may be the most important.

Although mycotoxicosis, as at present understood, is of limited direct importance in terms of human illness, the present state of knowledge suggests the advisability of vigilance in preventing the consumption of mould-infested food and further attention to safety (perhaps especially to the long-term safety) of the consumption of fungi deliberately used in food preparation. The report of an outbreak of hepatitis in India in 1974 affecting three hundred and ninety-seven persons of whom one hundred and six died gives point to this caution. They had all consumed chupatties and other foods made from maize flour heavily contaminated with *Aspergillus flavus.*

8.4.3. Food hygiene

From the foregoing brief outline of the principal causes of food poisoning in man, it will be seen that the primary source of infection of food is man and that the principal source of infection of man is food. Food hygiene is an attempt to interrupt the circle. Infection may begin in the domestic kitchen, in canteen, refectory, restaurant or other public eating place, in a food-processing plant, in vehicles or containers used in transit and even on the farm where the food was originally grown. Another factor is the relative harmlessness of the initial presence of a few food-poisoning organisms in a given sample of food but the great dangers associated with their growth and development in that food.

If food hygiene is a series of steps aimed at breaking a circle of infection, it properly starts with the food handler, whether in factory, shop or kitchen. Most infections are of intestinal origin and are spread from human faeces. Even the most fastidious may be unaware of the fact that lavatory paper is sufficiently porous to allow the transmission of micro-organisms to hands. (Try placing a piece on a normal inking pad and pressing the tips of your fingers against it.) Hands should therefore be washed with care after passing urine or stools. Observance of this simple precaution would probably do more to reducing the incidence of food poisoning in countries with a clean water supply than any other single factor. If it is important for the general population, it is essential for the food handler. Infected cuts, boils and pimples are also reservoirs of contamination. Those suffering from them should not be permitted to work in situations where they come into direct contact with foodstuffs especially where these are to be eaten without further cooking (for example, cream cakes and cold meats). Clothes may also transmit infection. External garments should be designed for comfort and for ease of cleaning and should be changed before contamination builds up. In some circumstances, such as are involved in the manufacture of whipping creams for the bakery trade, standards of hygiene as strict as for hospital operating theatres may be necessary, with access restricted to trained persons and changes of protective clothing on each ingress to and egress from the restricted area. Such situations are rare and the precautions are costly. In most circumstances, common sense must moderate caution. A worker in a warehouse receiving potatoes, turnips and carrots direct from the farm cannot keep exposed skin and clothing as clean as is essential for one handling cooked meats.

Attention to high standards of personal hygiene correlates with the provision of facilites to make this possible. Adequate toilet facilities with associated wash-hand basins provided with hot and cold water, soap and clean nail-brushes arc basic essentials. Particular care is needed in the provision of towels. There is little point in washing hands if they are subsequently dried on a grubby communal towel. Toilet facilities should not open directly into a work room and they must always be used exclusively for their designated purpose. In a busy pub it is by no means uncommon to see canned or bottled drinks stacked in a spare corner of a washroom. Since the drinks are sealed in impervious containers the practice seems harmless enough until one reflects on the people who drink straight from the can or bottle.

Even with the most dedicated standards of hygiene, the healthy carrier of an infection always presents a risk which can only be avoided by his or her identification. Detection is usually beyond the resource of an employer, and the local health authorities should be consulted in the event of doubt. However, the employer can observe mild respiratory infections (colds, sore throats etc.) and take steps to arrange the work of the person concerned so as to prevent sneezing or coughing over foods to be consumed without further cooking.

If the excellence of the hygiene standards of the food handler is the first line of defence against outbreaks of food poisoning, the back-up must lie in the standards of construction and layout of his place of work and in the quality of the fixed and movable equipment with which it is supplied. When a new restaurant or factory is designed it is easy to apply well-tried principles—a supply of clean air and water and an internal structure planned for ease of cleaning. To see to this is part of the professional expertise of the architect who has, or should have, an abundance of published information to draw on as well as national and local planning and hygiene regulations to satisfy before he proceeds to build. In practice existing buildings sometimes have to be adapted for food use and this may give rise to difficulties. The following are some of the more important points to take care of.

Ceilings, walls and floors must be finished with easily cleaned, impervious or non-porous materials which will not provide harbourage for dust or micro-organisms. Floors require special attention depending on whether they are to serve dry or wet processes. If the latter they must be carefully graded to drains to prevent lying water, and the drains themselves designed to ensure ease of cleaning. Account must also be taken if the floor is to sustain heavy trucking loads. Any surface breakages can become dangerous reservoirs of infection. For walls, tiles provide a good answer and, although relatively expensive, their durability, ease of cleaning and long life help to compensate for the initial cost. Service pipes and ducting placed high on walls accumulate infected dust which may be dislodged by draughts or vibration on to the foodstuffs in process below. Care at the initial planning stage can eliminate this hazard. Wood has unique properties and it has been used for food tables and utensils for centuries but it is porous and provides good harbourage for bacteria. Avoid it if possible. Take specialist advice if food is likely to come into contact with painted surfaces. Finally rodent-proofing and bird-proofing are essential. They can act as bacterial carriers, and foul working areas with droppings.

The selection and layout of equipment is a specialist topic and one which requires the professional advice of an engineer. However, few engineers have microbiological training, and while they understand the mechanical requirements and have an appreciation of the material specifications needed to prevent corrosion, they seldom understand hygiene problems. To resolve these in a practical way requires sympathetic understanding between engineer and food scientist or technologist, a situation catalysed by each having a minor understanding of the other's discipline and technical vocabulary. Most training courses for food scientists and technologists contain some basic engineering principles—heat and mass transfer, distillation, filtration and the like—but seldom do engineers have opportunity to study the elements of microbiology. There is, therefore, a risk that the selection and installation of equipment may reflect working convenience and mechanical effectiveness without equal consideration being given to ease of cleaning both in machine design and in its layout. Cleaning (and, where necessary, sterilising) of equipment is a constant and costly chore. Care at the design stage will raise standards and reduce later operating costs.

Food processing machinery takes such diverse forms that a text of this sort can only suggest key factors of general application. They may be summed up as follows.

1. All machine surfaces coming into physical contact with the food being processed are to be made from corrosion-resistant materials.
2. All such surfaces are to be smooth to avoid hidden foci of infection.
3. All such surfaces are to be accessible for cleaning purposes.
4. If immediate accessibility is not possible for mechanical reasons, ease of dismantling to provide access is essential.
5. Because of the nature of many foodstuffs, avoidance of spillage may sometimes be virtually impossible. It is therefore important to ensure the absence of dead spaces in which spilled food can accumulate and which can act as continuing foci of infection.
6. Proper attention must be given to machine bases, especially where a heavy machine is bolted to a floor. Without care, and especially in wet or

moist working conditions, hollow sections in the base of machines can harbour debris, insects and the like.

7. Many food-processing operations involve the use of steam or the heating of open vessels containing liquid foods. Proper ventilation to extract steam is essential. Condensation on roofs and walls forms heavily contaminated drops, which may fall into and contaminate the material being processed. Furthermore, unventilated steamy atmospheres make for bad working conditions, and those reduce the vigilance and cooperation of the work force.

8. Cleaning systems should not be an afterthought but should be planned from the outset as part of the plant assembly. Services which may be required for the cleaning process (e.g. power and steam hoses) are thus brought to the point of use. The more convenient the layout and the more accessible the services, the more likely it is that cleaning will be properly performed.

9. Cleaning normally has to be carried out when the plant is not in operation. Under conditions of pressure (for example, pea-processing lines at the peak of the harvest season) it is tempting to reduce cleaning time to take production overload. Cleaning time must be built into the production plan and properly scheduled and supervised.

10. A regular system of inspection and bacteriological checking helps to ensure that the procedures adopted are effective.

11. Adequate and convenient storage must be provided for cleaning equipment, detergents and disinfectants.

In addition to machines and working surfaces in factory or retail outlets, utensils have to be used. These range from beer mugs in pubs, cutlery and crockery in restaurants, to trolleys and containers in factories. Whether large or small, proper provision for cleaning what are often regarded as unimportant auxiliaries matters as much (or sometimes more) than cleaning the main machines in a production line. Again the provision of proper and convenient washing facilities for movable materials is essential, and these must be quite separate from the wash-hand basins provided for personal hygiene. It is obvious that such facilities require adequate hot and cold water supplies, and present regulations in the United Kingdom insist that the hot water should be at a temperature of not less than 170°F (i.e. 77°C), which is sufficiently hot to kill vegetative bacterial cells.

Having paid attention to the design of factory buildings, the layout and ease of cleaning of machines and the provison of proper washing arrangements for utensils, there is one further factor which is a function of management and which is often forgotten or ignored. It is orderliness.

There is one further factor which is a function of management and which

is often forgotten or ignored. It is orderliness. I have repeated the sentences because of their importance.

Human beings tend to be slothful rather than energetic, lazy rather than industrious, careless rather than careful and untidy rather than tidy. Opposing these tendencies are other human characteristics of intelligence, curiosity and awareness of the dangers of laziness, carelessness and untidiness. When a group of people work together, whether as a football team or a working gang in a factory, the emphasis of these individual characteristics is summed up through the individuals into a group character. When laziness predominates, we get slipshod play on the football field and an untidy and unhygienic sector in the factory. However, if the same team has a good captain or work gang a good foreman who through exhortation and personal example shows the advantages of tidy play on the field or a tidy environment in the factory, standards quickly and almost automatically rise. In the factory one of these to rise is hygiene.

However, the factory charge-hand cannot set a desired example unless higher levels of management provide the ways and means. Good standards of hygiene are impossible without effective organisation of materials. The following are the key points.

1. A place for everything and everything in its place. Proper and convenient storage must be provided for tools, trolleys, storage bins, cleaning materials and all movable equipment. Floor surfaces must be kept clear as also must passages between work rooms.
2. Foodstuffs in process must not be allowed to accumulate in a disorderly fashion at intermediate stages of production. Ingredient preparation and weighing (such as salt and spices or flavouring materials) are often regarded as minor functions and carried out in a sloppy careless fashion. This stage must be properly planned, systematised and recorded.
3. Materials should only be drawn from the main storerooms or factory warehouses as required. It should be the general rule that not more than one day's supply be held in a department or work-room at any one time.
4. Higher management must pay most careful attention to the operation of the main factory store-rooms. Tidiness further downstream is impossible without a well organised and well-run stores system.
5. Employees should not bring outer clothing into workrooms, nor should they take meals in these rooms.
6. Proper waste-bins must be provided at convenient points within workrooms.

Orderliness is a management objective which produces good working conditions, reduces worker fatigue and improves the working environment. These are appropriate to the civilised working of an organisation, but this is not our immediate concern. Orderliness and good hygienic practice go together. Good standards of hygiene are impossible in untidy cluttered workplaces. If workpeople are well-groomed, clean and tidy (and they will be if given the right conditions) and if the place of work is bright, tidy and easy to clean, management has laid the foundations for good hygienic practices, and for the elimination of the risk of that curse of contemporary society, food poisoning.

8.5. Food quality assessment

8.5.1. Introduction

The nutritional quality of foods has been dealt with in Chapter 1. As has often been pointed out, unconsumed food has no nutritional value whatsoever. In dealing with the detail of the range of problems which confront him, it is easy for the food scientist to allow the human factors to be forgotten in his search for technical solutions. The consumption of food involves more than the satisfaction of nutritional needs. It is the basis of much of our patterns of social behaviour. Some the factors involved are as follows.

a) Meals are, or should be, a major source of enjoyment and refreshment.

b) Pleasure is enhanced if eating is also a social event. To throw a party without food and drink is unthinkable.

c) Family meals reinforce the importance of the family as the basic social unit of structure in society.

d) Food consumption provides appropriate periods of rest and relaxation by dividing the work periods of the day.

e) Food consumption has long had a religious significance as illustrated by the feasts and fasts enjoined by the major religions of the world. It thus contributes in another form to bringing family units into larger groupings which share common beliefs and attitudes.

f) Food consumption has a cultural significance in terms of the cuisines associated with different ethnic, national and regional resources and traditions.

g) Food consumption has economic significance in terms of local, national and international trade.

h) Food consumption has a technological significance in terms of the inventiveness generated by the need to satisfy hunger.

i) Food consumption has a political significance in terms of the instability created between nations by threats to their food supply.

All these, and many others, lie behind the simple and familiar busines of sitting down to a meal. They affect the work of the food scientist because in the process of consuming that meal the meal is judged, consciously or unconsciously for its quality. The matter is complicated by the fact that ideas of food quality vary from person to person, from country to country and from one foodstuff to another. The assessment of food quality therefore varies with the situation. The underlying factors remain constant. In an immediate way these were neatly summed up by a student who, in answering a question on food quality, wrote 'food tastes better if it is prepared by someone you love'.

There is a big gap between this statement and the daily business of the food scientist who is more commonly concerned with the bulk handling and processing of food than with the personalised concerns of the domestic kitchen. A good mixed diet calls for an annual consumption of around 500 kg of food per person (wet weight as consumed). Thus from a national point of view 500,000 tonnes of food are required per million inhabitants. In industrialised societies with high urban concentrations and low rural population densities the direct transfer of food from farm to table is rarely possible. The farm product is merely a raw material which must be processed and preserved in a wide variety of ways before it reaches the complex distribution system which eventually brings into shop and supermarket the huge range of foods from all corners of the globe which provide the wide range of choice now available, a range which has greatly increased during the past twenty-five years. The maintenance of quality standards aimed at retaining both attractiveness and nutritional value throughout this complex chain of events is a major activity of scientists working in the food field. Techniques for doing this have been developed gradually over the past hundred years. During the past quarter century they have become systematic as growing knowledge of the properties of foodstuffs has been associated with improvements in methods of measurement.

8.5.2. What is meant by food quality?

The simple answer to this question is 'those characteristics of a food which make it agreeable to the consumer'. This raises further and related questions. Which food? Which consumer? What price? For what purpose? Consumers' tastes differ even for a single food like an apple—some enjoy Cox's Pippins, others prefer MacIntosh Reds or Golden Delicious when the apple is to be eaten raw. If it is to be cooked, many would prefer a Bramley Seedling. The preference is further complicated by price. Many people say they prefer salmon to cod but eat cod because it is cheaper. We can, therefore, see that the aim is produce a product whose quality characteristics match the price at which it is to be sold: that is to say they match the needs of that sector of the market at which it is aimed. In practice therefore the assessment of food quality must be appropriate to the market aimed at.

Markets are socially determined and are subject to the influences of social mixing and social change. The demands of the caviare and champagne market are different from those of the pie and a pint pub. In making quality assessments like must be compared with like. Within a given country, many local variations in eating patterns are readily observed. To a foreigner the concept of 'French cuisine', for example, raises a generalised mental image, while to a Frenchman 'French cuisine' is an amalgam of his experience of a range of local cooking patterns often partly based on local crops.

The origins of food habits and the criteria applied by individuals in making a choice of food are perhaps studies for disciplines other than those normally used by food scientists, but a few general comments may form a useful background at this stage. The wide range of human likes and dislikes of given foodstuffs is a matter of common observation. The food scientist can make measurements of colour and texture of foodstuffs with the laboratory tools of his trade. These measurements enable him to make comparisons of one food with another in objective terms which he can express in numerical form. He can analyse food for the small amounts of chemical substances which confer flavour and in some cases quantify these numerically. This information tells him nothing about whether one individual will like or dislike the food in question. However, if he compares these numbers with the likes and dislikes of a sufficiently large and randomly selected group of consumers he can arrive at reasonable conclusions about the relationships between the laboratory-generated numbers and the market's (i.e. his customer's) likes and dislikes. Thus he may be able to say (for example) that nine people out of ten like strawberry jam made by a certain formula and process. By pushing the analysis further he may even be able to find out the main reasons why some people like this particular strawberry jam and others do not, and information of this sort is clearly of great commercial value. No one wishes to invest money in designing, building and equipping a factory and in marketing its products if the products themselves do not appeal to the market sector at which they are aimed.

The nature of 'liking' is subtle. In addition to those immediate sensory reactions to appearance, smell, taste and texture, most observers would agree that other factors affect food selection.

1. Social background and upbringing. Food likes and dislikes tend to be acquired in childhood and adolescence. These are retained in later life, although somewhat modified by experience.

2. Primitive or sophisticated concepts of hygiene. Badly chosen food can give rise to sickness. Food must be 'clean'. Pure food is desirable. Food must be 'safe'. A mother cooking for a family is more likely to

ensure that food is clean, pure and safe than a stranger doing the same job. (These, of course, are examples of attitudes and not necessarily of fact!)

3. Social custom. An Eskimo is not likely to take kindly to rice. Raw fish or raw meat are regarded as unacceptable in some cultures and prized in others.

4. Religious requirements. Muslim and Jewish faiths forbid the consumption of pork and both require special conditions when animals are killed for food. Food prohibitions are associated with most major religions.

5. Mood. In a given individual, choice will vary from day to day depending on physiological or emotional state. 'I don't feel like . . . today' is the common expression of this experience.

6. Weather or climate. In hot weather or on a change from a temperate to a tropical climate patterns of food selection change.

7. Pregnancy. During pregnancy food habits sometimes change not merely in the amount but also in the kinds of food eaten.

Other factors (for example, sickness, old age, adolescence, or simple boredom) may also come into play. Add all these together and individual behaviour may often appear more capricious than rational to a detached observer. On the other hand, differences level out when group behaviour is observed and the bigger the group the greater the levelling effect.

By its nature, food processing tends to be a large-scale activity with a group market of substantial size. The product design and quality control functions (our objective is to make the best possible tomato soup, or whatever!) aimed at a mass market approach their task from a quite different viewpoint to that of the chef. Even in a large and expensive restaurant, the chef prepares food for a quite small number of customers to whom he is going to offer a range of dishes from which they can choose. In contrast, the food processor normally offers a single product of its type. His design and objective are to get the given product formulation (say tomato soup) which appeals to the largest possible number of people and to sustain this quality despite changes in the raw materials which he uses.

On a given day and with a given menu the chef will do his artistic best to produce his concept of say, *sole à la Nantua*, but the dish as he prepares it will vary somewhat from day to day. The processed food product must not vary in a way the customer can detect. An expectation is created by the first purchase which, if not sustained on subsequent purchases, will result in loss of business. The approach of the food laboratory is thus quite different to that of the kitchen, and food quality concepts of a process development laboratory will not necessarily harmonise with those of the chef.

8.5.3. The techniques of food quality assessment.

The general approach to assessing food quality is to obtain detailed information about factors affecting product quality, preferably in numerical form, and then collate this information to produce a picture of the product which can be used on a week-to-week or month-to-month basis of comparison. Once such a standard has been set, the range of variation observed during production operations can be observed, and limits set as to what is acceptable. Some variation is unavoidable. If its range is less than the consumer is likely to observe, it does not matter. Tolerance will therefore be based on the limits set by this range.

Further restrictions on the product, the container in which it is sold and the label used to describe it, are placed by legislation established by benevolent governments around the world to protect their peoples from fraud or from the sale of food containing substances dangerous or potentially dangerous to health. To ensure that a given good product complies with legal requirements is a normal part of the design specification and its implementation a function of the quality control operation.

The basis of measurement can be summed up as follows.

1. Measurements on the container (can, bag, box, wrapper etc.) which will establish whether its functional properties are up to specification.
2. Measurements on the raw materials from which the product is to be produced to ensure that they are appropriate to the function for which they are intended.
3. Measurements on the processing operation to ensure that it has been properly performed.
4. Measurements on the finished product before it leaves the plant to ensure that it falls within the limits laid down by the product specification.
5. Measurements on the finished product at the point of sale to ensure that the product has not deteriorated during distribution and storage.

If the first three sets of measurements are carried out systematically and thoroughly, and if the product and its container have been properly designed for the expected storage life, the need for extensive sampling and testing at stages 4 and 5 is greatly reduced.

A complete set of measurements made at any one time is like a single frame in a cine film. A sequence of measurements made at appropriate intervals gives time-related information. It tells, for example, whether the weight filled by a filling machine drifts upwards or downwards or remains constant as the working day progresses. On a longer time-base it tells us how the quality of a processed crop varies through the season of harvest. Other variations may occur in such factors as physical dimensions, temperature, processing times, particle sizes, pressures and the like, any one of which may start a set of changes leading to deterioration of the product quality. The assessment and control of quality involves measurements made consecutively against a time-base from which a product image emerges varying fractionally with time. Corrective steps may be taken if and as the image distorts.

The techniques of measurement are matters which depend on the nature of the product and the function to be measured. Any kind of test may be used: physical, chemical, biochemical, microbiological and organoleptic methods are all commonly employed. Raw material examination may include such simple criteria as freedom from defects (for example, 'less than 1 per cent of peas showing insect damage') or the chemical analysis (for example, the iodine value, saponification value, refractive index and free fatty acid figure) which might bc rcquired when considering deliveries of oils and fats. Water is an important raw material and microbiological controls may be called for. Meat, fish and milk are also ready media for micro-organisms and similarly their use may necessitate microbiological controls. These kinds of techniques will already be familiar to the reader and many excellent works are available if detailed technical information on how to conduct them is needed.

Organoleptic methods may not be familiar and they therefore require brief consideration here. The term is simply a technical one referring to the human sense of taste and smell, but it is now used to describe quantitative subjective responses of human tasters to a food situation. In certain industries the role of the expert taster is traditional. Tea blending, whisky blending and wine merchandising have long depended on the skills of professional tasters. Such people are (or claim to be) sensitive to an unusual degree to the flavour impressions in the products in which they are expert. These natural skills are improved by long experience which probably does nothing to increase sensitivity of impression, but does systematise the sensory information at their disposal and may also link it to a descriptive vocabulary. Such individuals are useful to the trades they serve because their ability to discriminate is greater than that of the average customer for the product. However, prima donnas have off-days as have expert tasters, and their mistakes can be very costly. Moreover replacing them when they retire may present formidable problems. Even at their best and most useful, their data tend to be of the pass–fail type, and graded information is of greater practical value. A few have been self-appointed experts whose real skills were not as great as they claimed. For these and other reasons, professional tasters of the traditional type are less frequently used now than in the past. Their function has been taken over by tasting panels.

Tasting panels are simply groups of people brought together in controlled surroundings to taste foods according to a predetermined plan and to record their results in a systematic way which can be analysed by statistical methods if required. They are composed of normal individuals rather than specially gifted ones and therefore tend to react towards food situations in a way which more typically reflects the reactions of normal customers. If a member of the panel falls ill or resigns he or she can be comparatively easily replaced, off-days of individuals are sometimes detectable in the panel situation and, even if not, their effects play a relatively minor part in the overall results. Most important of all, the data generated are usually in numerical form which can be treated statistically and tested for reliability.

Tasting panels can be used for a variety of purposes and their composition and *modus operandi* will depend on the objective to be achieved. They have been extensively used at research level in attempts to link chemical or microbiological changes in stored food products with deterioration from the consumers viewpoint. They have been used in product development to assess progress in product formulations. They have been used in product appraisal exercises in comparing the characteristics of different market brands. Perhaps the most common use of all is in the day-to-day control of mass-produced branded foodstuffs. It is this use with which we are concerned here.

The design of the panel and the selection of its membership is a matter for careful thought. Before proceeding, it is essential to agree on and commit to paper a statement of its purpose and function. It is very easy to start with confused ideas on this and unless these are clarified before work begins on panel selection the outcome will be equally confused.

The size of the panel and its membership will depend on its function and purpose. For day-to-day quality control purposes in a production situation, the panel will probably meet every working day. Daily attendance may reduce interest, and boredom brings carelessness. Account has to be taken of holidays and absence for reasons such as sickness or business travel. In the light of considerations such as these, it may be decided to have a membership of say, twelve (who will train together during the formation stages) with the intention of using a rota of five or six for any one panel session. It is also desirable to have a panel organiser, responsible for calling meetings, selecting and preparing the samples to be presented to the panel and analysing the data generated. Normally the organiser should not take part in tasting, but good rules often have exceptions and situations do arise when it is proper for the panel organiser to participate.

An example of such a circumstance is the small panel involving, say, four members who have each a long experience of the product in question. Such panels are cheap, speedy and effective in operation and are much easier to organise than larger panels. Their disadvantage is a tendency to bias. In practice the membership is often drawn from the ranks of middle or senior management, since it is there that good product experience is to be found. As a result, on many working days one or another member will be called away on other duties (tasting panels tend to be of low priority if a strike is threatened or raw material supplies are at risk!). Full attendance at meetings may therefore be less frequent than desirable, and the outcome may be data of doubtful statistical validity. Despite this, experience has shown that a well-chosen panel of this type can operate satisfactorily and perform their designated function inexpensively and accurately providing they are conscious of their own limitations. One useful safeguard is that members agree to seek further opinions outside the normal panel in cases where there is disagreement within the panel.

The selection of members for whatever type of panel has been much written about. The more important criteria are considered here but there may be others applying to specific cases.

Panel members should be of equable temperament. Strongly opinioned and aggressive individuals on the one hand or highly strung and affective individuals on the other are to be avoided if possible.

Since taste sensitivity varies from one person to another, both in overall terms and in relationship to specific factors such as bitterness, and since a proportion of a given population is taste-blind to certain flavour impressions, it is useful to check that candidates for panel membership have reasonably sensitive palates. One way to do this is by threshold tests. The candidate is offered very dilute solutions of a known flavouring substance in order of increasing concentration and asked to report on the level at which the flavour is first detected. As Gridgeman has pointed out, a person who can detect sweetness in a 1.5 per cent solution of sugar, saltiness in a 0.15 per cent solution of sodium chloride, sourness in a 0.06 per cent solution of citric acid and bitterness in a 0.005 per cent solution of quinine sulphate is likely to be good at describing his sensory impressions. For sensitivity, the person who can rank sucrose solutions of 7.5 per cent, 10 per cent, 12.5 per cent and 15 per cent in ascending order has a good palate. (Gridgeman, N. in *Quality Control in the Food Industry*, Herschdoerfer, S. M. ed., vol. 1, pp. 245–247).

The odds of getting this sequence right by chance are 24 to 1 against. Tests such as these, and others for the sense of smell, may be used to ensure that the panel membership has appropriate organoleptic sensitivity.

It is essential also that panel members are interested in the work of the panel. For this reason they should be selected from volunteers. In general the inherent interest of the work with a properly motivated panel is reward in itself, but if membership involves a modest status and small privileges, the additional incentive is useful.

Normally the panel should consist of both men and women in about equal numbers but this is not a strict ruling and there is little evidence to suggest that there are substantial differences between the sexes in terms of flavour and odour sensitivity. The evidence on the influence of smoking and wearing dentures suggests that these have less influence on organolepsis than some writers imply. There seems to be agreement that the senses of taste and smell become less acute with age. However, this general statement hides the fact that some elderly men and women retain their taste and odour perceptiveness to a marked degree and are just as sensitive as the young.

If reliable results are to be obtained, taste panel work must be carried out in appropriate surroundings, in a quiet room specially set aside for the purpose, free from cooking and other smells, the food to be presented being prepared in an adjoining kitchen. The lighting must be constant when colour assessments are to be made. Normally, members should not consult with one another until they have completed and signed their score cards. Two types of situation are in common use. In one the prepared samples are placed on a central table around which the panel members gather to make their judgements. In the second a range of booths (like voting booths) are set along one wall, each booth being provided with controlled lighting, water to rinse the mouth between tests and a hatch to the adjoining kitchen, through which the test samples are passed. In general the second method is to be preferred since it ensures privacy of judgement. Tasting should normally take place midway between meals when neither hungry nor sated. These precautions are necessary in the interests of excluding extraneous influences. Nevertheless the controlled tasting situation is far removed from the circumstances of normal eating when mixtures of foods are consumed in social surroundings.

When a new panel has been selected for a defined set of purposes it is necessary to devise a common vocabulary to describe the impressions received and to regularise them by an agreed system of numbers. Suppose, for example, that the panel is given the comparatively straightforward task of monitoring product quality of peas from a freezing plant to an agreed sampling plan. A set procedure for cooking the peas has been established and a scoring or grading system has been prepared along the following lines.

Colour

Grade 4. A fresh, bright-green, uniform pea colour equivalent to freshly picked, freshly cooked garden peas. Free from discoloured or insect-damaged peas.

Grade 3. A good green colour but some variation in shade detected. Not more than 2 discoloured or insect-damaged peas per 100-g serving.

Grade 2. An average green colour with up to 10 per cent colour variation. Not more than five discoloured or insect-damaged peas per 100-g serving.

Grade 1. Poor colour with up to 20% colour variation. Not more than ten discoloured or insect damaged peas per 100-g serving.

Flavour

Grade 4. Sweet, full, natural pea flavour equivalent to freshly picked freshly cooked garden peas. Free from all traces of off-flavours.

Grade 3. Sweet, natural pea flavour.

Grade 2. Less sweet but natural pea flavour with detectable 'hay-like' off-flavour. No trace of mouldy off-flavour.

Grade 1. Starchy flavour. 'Hay-like' or mouldy off-flavours noticeable.

Texture

Grade 4. Tender cotyledons, soft skins. Not more than 2 per cent of split or loose skins, equivalent to freshly picked freshly cooked garden peas. Free from all traces of mealiness or toughness.

Grade 3. Tender cotyledons, some skin toughness acceptable. Not more than 5 per cent of split or loose skins. Free from mealiness.

Grade 2. Detectable mealiness, and relatively tough skins. Not more than 10 per cent of split or loose skins.

Grade 1. Mealy overmature peas of tough texture.

The reader will note that the scheme outlined is really a system designed to define commercial grades of quality. Our quality control panel, given this scheme, might well decide to use it as a basis for scoring, using a card drawn up as follows, and showing a set of tasting data.

Sample code	Colour (4-max.)	Flavour (4-max.)	Texture (4-max.)	Total (12-max.)
PF1/30/6	3.0	2.8	3.2	9.0
PF2/30/6	2.7	2.8	2.9	8.4
PF3/30/6	2.5	2.6	2.6	7.7
PF4/30/6	2.2	2.4	2.6	7.2
PF5/30/6	3.1	3.1	3.3	9.5

Date 1st July *Signed* John Black

The card reports one panel member's marks for a set of five samples scored on the 1st July on samples processed on the 30th June. It will be noted that none of them fall exactly within the category descriptions of the grade specifications and the scorer has used decimal fractions to indicate where between each original grade he thinks the product lies for each characteristic. In fact, the marking range is not out of four but out of forty, the number 4.0 having become the unattainable ideal of the perfect frozen pea and zero representing a concept of the unimaginably bad. In practical situations, unattainable ideals do not exist and unimaginably bad consignments would not get through the factory gates. As a result marks above 3.5 or below 1.5 are seldom recorded, and the effective marking range has become 20.

Over several weeks or several seasons a panel operating such a system ceases to need the written word as a guide. A mental picture, the integration of a large number of previous panel experiences, becomes the real reference standard. A new member joining the panel acquires, in time, a similar mental image which gives continuity over the years to the panel's work as external circumstances change. To say this is not to underestimate the significance of the written grade specification, which will be referred to when panel members disagree and which soon becomes part of the overall mental picture.

At a given panel session involving six members, the six score cards will generate twenty-four pieces of information in the form of figures for each sample examined. Since personal judgements are involved, the figures are not expected to agree exactly. As a simple first approach, the six sets of figures may simply be averaged to make the data manageable, and during the early training stages of a new panel this is a useful procedure, especially if the means are afterwards shown to individual members who can thereby see whether or not they greatly differ in judgement from the averaged opinion. After the panel has settled down to regular meetings, standard deviations can be calculated on individual samples and these will give an index of the range of variation of opinion between individual scores for a given character. Other statistical refinements can be applied depending on the kinds of questions one wishes to ask on the further significance of the data generated by the panel. Geographically separated processing plants producing the same product to the same specification and using local tasting panels for control purposes may wish to correlate their results to ensure that each panel is marking to the same standards. The procedures to be used are beyond the scope of this book and reference may be made to the sources quoted at the end of this chapter.

One reservation about the nature of the data generated by schemes such as these is whether there is logical justification for using numbers and treating them arithmetically. We cannot be sure that the difference between a score of say 2.7 and 2.9 for flavour represents the same change in flavour value as that between say 3.3 and 3.5. For practical purposes we assume that this is so but the proof of the matter is not simple, and this factor should be remembered when interpreting either directly or statistically.

In practice, and operating within defined and controlled conditions, experience-based pass and fail limits are applied and appear to work quite satisfactorily despite this reservation. Going back to the set of data already considered, let us assume that the peas being tested have three possible uses. The top quality is to be used for own-brand retail sale. The second quality is for high-grade catering outlets and the remainder will be diverted to an in-plant canning line where over-mature and less sweet peas are quite acceptable, since the sterilising process involves temperatures which soften the harder peas.

The specification, for example, might define the own-brand retail grade as samples scoring a minimum of 2.8 on each character and a total mark of not less than 9.0. The catering pack might operate at a minimum of 2.4 on each grade and a total of not less than 7.5. Members of the panel would know this and John Black's score card can be interpreted to mean that he considered only the first and last sample as being good enough for the top quality pack, he thought the second and third satisfactory for a catering pack and he rated the fourth as being best used for canning. But from the dates it is clear that the samples were tested the day after passing through the factory when it is too late to divert part of the incoming product to a canning line. Furthermore, depending on the type of freezing plant used, it may be inconvenient and expensive or even impossible to reallocate stock to different outlets after processing, especially as retail and catering grades are normally distributed in different sized containers.

In addition to taste-panel criteria each grade will have specified physical, chemical and perhaps microbiological standards to meet. Again, only limited laboratory testing is possible when a production run is in progress and some aspects of it may not be completed for a day or even several days afterwards. Only when all this information has been obtained can final disposal decisions be made. Immediate decisions are required on plant and these are the function of the process control system. To consider it and how it relates to what has been discussed under this heading of food quality assessment takes us from food science to food technology and beyond the intent of this volume. It is sufficient to say here that rapid in-line techniques are usually used to provide provisional quality ratings for plant use and these are later subject to more precise laboratory evaluation.

The quality information as finally evaluated can be used in ways other than the direct control function described. With a product such as peas, season-to-season comparisons may be needed for planning purposes, comparisons between early and late varieties may be needed when sequential ripening systems are used to smooth raw material flow during the peak season, comparisons of quality of the same varieties grown on different soil types, and trials aimed at evaluating the processing quality of new varieties may all be essential for the future well-being of the processing company.

Although peas have been chosen as an illustrative example, the underlying concepts they illustrate can be adapted for any other type of foodstuff.

Further reference

Lipid Biochemistry (1971) Girr, M. I. and James, A. T., Chapman and Hall, London.

Margarine (1969) van Stuyvenberg, J. H., Liverpool University Press.

Introduction to the Biochemistry of Foods (1976) Berk, Z., 2nd edn, pp. 169–182 (for oils and fats) and pp. 149–167 (for non-enzymic browning), Elsevier, Oxford and New York.

The Biochemistry of Foods (1971) Eskin, N. A. M., Henderson, H. M. and Townsend, R. J. (browning reactions, pp. 69–103), Academic Press, New York and London.

Food Poisoning and Food Hygiene (1978) Hobbs, Betty C. and Gilbert, J. 4th edn, Edward Arnold, London.

Food Borne Infections and Intoxications (1969) Riemann, H. (ed.) Food Monograph Series, Academic Press.

Microorganisms in Foods (1968) Thatcher, F. S. and Clark, D. S. (eds) University of Toronto Press.

Bacterial Food Poisoning (1969) Taylor, J., The Royal Society of Health, London. (For *Salmonella* infections.)

Modern Food Microbiology (1978) Jay, J. M., 2nd edn, van Nostrand, New York and London.

Quality Control in the Food Industry (1967) Herschdoerfer, S. M., (ed.), vol. 1, Academic Press, New York and London.

Index

Numbers in bold type refer to main sections